Squid

Animal
Series editor: Jonathan Burt

Already published

Albatross Graham Barwell · *Ant* Charlotte Sleigh · *Ape* John Sorenson · *Badger* Daniel Heath Justice
Bat Tessa Laird · *Bear* Robert E. Bieder · *Beaver* Rachel Poliquin · *Bedbug* Klaus Reinhardt
Bee Claire Preston · *Beetle* Adam Dodd · *Bison* Desmond Morris · *Camel* Robert Irwin
Cat Katharine M. Rogers · *Chicken* Annie Potts · *Cockroach* Marion Copeland · *Cow* Hannah Velten
Crocodile Dan Wylie · *Crow* Boria Sax · *Deer* John Fletcher · *Dog* Susan McHugh · *Dolphin* Alan Rauch
Donkey Jill Bough · *Duck* Victoria de Rijke · *Eagle* Janine Rogers · *Eel* Richard Schweid
Elephant Dan Wylie · *Falcon* Helen Macdonald · *Flamingo* Caitlin R. Kight · *Fly* Steven Connor
Fox Martin Wallen · *Frog* Charlotte Sleigh · *Giraffe* Edgar Williams · *Goat* Joy Hinson
Goldfish Anna Marie Roos · *Gorilla* Ted Gott and Kathryn Weir · *Guinea Pig* Dorothy Yamamoto
Hare Simon Carnell · *Hedgehog* Hugh Warwick · *Hippopotamus* Edgar Williams · *Horse* Elaine Walker
Hyena Mikita Brottman · *Human* Amanda Rees and Charlotte Sleigh · *Jellyfish* Peter Williams
Kangaroo John Simons · *Kingfisher* Ildiko Szabo · *Leech* Robert G. W. Kirk and Neil Pemberton
Leopard Desmond Morris · *Lion* Deirdre Jackson · *Lizard* Boria Sax · *Llama* Helen Cowie
Lobster Richard J. Kin · *Mole* Steve Gronert Ellerhoff · *Monkey* Desmond Morris · *Moose* Kevin Jackson
Mosquito Richard Jones · *Moth* Matthew Gandy · *Mouse* Georgie Carroll · *Octopus* Richard Schweid
Ostrich Edgar Williams · *Otter* Daniel Allen · *Owl* Desmond Morris · *Oyster* Rebecca Stott
Parrot Paul Carter · *Peacock* Christine E. Jackson · *Pelican* Barbara Allen · *Penguin* Stephen Martin
Pig Brett Mizelle · *Pigeon* Barbara Allen · *Polar Bear* Margery Fee · *Rat* Jonathan Burt
Rhinoceros Kelly Enright · *Salmon* Peter Coates · *Sardine* Trevor Day · *Scorpion* Louise M. Pryke
Seal Victoria Dickenson · *Shark* Dean Crawford · *Sheep* Philip Armstrong · *Skunk* Alyce Miller
Snail Peter Williams · *Snake* Drake Stutesman · *Sparrow* Kim Todd · *Spider* Katarzyna and Sergiusz
Michalski · *Squid* Martin Wallen · *Swallow* Angela Turner · *Swan* Peter Young · *Tiger* Susie Green
Tortoise Peter Young · *Trout* James Owen · *Turtle* Louise M. Pryke · *Vulture* Thom van Dooren
Walrus John Miller and Louise Miller · *Wasp* Richard Jones · *Whale* Joe Roman
Wild Boar Dorothy Yamamoto · *Wolf* Garry Marvin · *Woodpecker* Gerard Gorman
Zebra Christopher Plumb and Samuel Shaw

Squid

Martin Wallen

REAKTION BOOKS

Published by
REAKTION BOOKS LTD
Unit 32, Waterside
44–48 Wharf Road
London N1 7UX, UK
www.reaktionbooks.co.uk

First published 2021
Copyright © Martin Wallen 2021

Printed and bound in India by Replika Press Pvt. Ltd

A catalogue record for this book is available from the British Library

ISBN 978 1 78914 334 8

Contents

1 Natural Histories from Aristotle to Steenstrup

Squids live for a brief period. After a year of predatory eating, they copulate with one or more other squids, expel their large masses of eggs, then die, sinking to the ocean floor, where their boneless bodies decompose, leaving little, if anything, to be fossilized. This crude summary of the squid life cycle serves over and over in the modern scientific studies to explain or justify the dearth of understanding about squid life. It appears as the explanation for ignorance about what a squid eats, and for uncertainty about the movements of various species through the seas and oceans. Such factual statements barely begin to approach the strange, even paradoxical interests squids have aroused in humans over the millennia.

Modern marine biology has drastically changed humans' understanding of squids. It has made us aware of a dizzying number of genera and species, and, with at least one species, that a creature many people had dismissed as fantasy does in fact exist. The simple factual account of squids – that they live fast, die young, have ten appendages and enormous eyes, and change colours continually – does not get at what 'squid' means. The various ways we study squids, represent them in the arts, exploit them and give them prominent roles in our nightmares all interact to shape the meaning of 'squid'. And these different ways of engaging with the idea and facts of squids have their own histories, which the following chapters will explore in turn.

Squid catching bait in the Maldives.

For as long as observers in the West have attempted to describe the natural world systematically – that is, since Aristotle – natural historians have expressed uncertainties about cephalopods. Octopuses, with their eight legs and no tentacles, swim in a clearly different category from the two decapods – cuttlefishes and squids – who add a pair of prey-seizing tentacles to their eight legs. The writers who come after Aristotle often use the names of these three cephalopod families interchangeably, reflecting a general uncertainty over which animal is which. That uncertainty becomes all the greater when translators with more linguistic training than scientific rely on informed guesswork in choosing the terms in their own language to correspond to those in the classical texts. Some of these terms come from legends of monsters, adding even more complications. The resulting tangle requires that we approach this long tradition of natural histories with something more than an informed skill at decoding what they mean. Along with occasional anxieties about sea monsters, all of these writers brought a love for the natural world and a desire to engage with their intellectual predecessors to the task of describing the creatures in the world, even those they had never personally encountered or those they feared.

It is this same combination of interest in the creatures we share the world with and admiration for the efforts by earlier writers to describe the world truthfully that I wish to bring to reading the natural histories, science and myths of squids. To put it simply, I believe that earlier non-scientific accounts of squids should be read as sincere attempts to identify these creatures accurately. Every era and every culture describes the world accurately in the terms available to them and in accord with their beliefs and customs. As we follow the long sequence of natural historians who followed from Aristotle in trying to understand what a squid looks like, what it does and how it is related to other creatures, we can see these informed observers bringing their various cultural assumptions along with them. These assumptions reflect very different attitudes towards non-human animals, and especially those in the seas. Even as sea creatures have been caught and served as ordinary food and rare delicacies by people along the shore, they have also inspired anxieties over their habitation in the watery deeps. All these cultural attitudes have shaped the descriptions and explanations of squids and even laid the foundation for modern science, as well as for literature and the arts. In order to set the studies of squids conducted by modern marine biologists in their historical context, I shall follow the successive efforts over two millennia to explain what a squid is. Taken together, this series of descriptive explanations by natural historians makes for a revealing study in the way caring and informed people create their own uncertainties and anxieties.

Aristotle made the first contributions to Western natural history, for his pre-Socratic predecessors tended to focus on the divine elements, to ask how they originated and formed the world. Aristotle turned his attention to the beings currently inhabiting the various elements, detailing their differences, and always in reference to the element that gave them their distinctness. The place where a creature lives determines the substance of which it is made, so that solid creatures live on land, hot creatures live in the bright air, and soft, fluid creatures populate the waters.[1] As inhabitants of the water, and even of the sea away from land, squids would be watery in their 'matter'; as he looks to the cephalopods generally and squids in particular, Aristotle analyses their physical make-up and their behaviour (confined mostly to what they eat and how they move and reproduce) in terms of their wateriness. He refers to watery squids with the term *malakion*, a word that emphasizes softness, and which might be rendered 'squishy

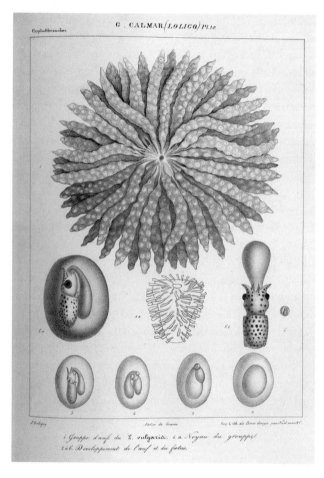

ones'. In addition, he categorizes squids – as with all cephalopods – as bloodless animals, a decision which gets him into some difficulty, as the gods whose veins carry ichor are also 'bloodless', so that Aristotle cannot figure out what to call the sub-lunar counterpart of blood.[2] Squishy and bloodless, cephalopods reflect the environment through which they move, and are thus cold, with almost no earthy part.[3] Cold and soft, these animals hold no identity with the terrestrial world of solids and warmth or with the heavenly world of ambrosia. Like the other three categories of bloodless animals (crustaceans, insects and testaceans), cephalopods lay 'imperfect' eggs, meaning that, once laid, the eggs continue to grow, unlike the 'perfect' eggs of birds and reptiles. The other bloodless animals all carry an earthy element in their shells, while the squishies lay squishy, watery eggs. To protect their eggs, cephalopods cover them with a sticky substance, which reflects the stickiness of their bodies.[4] Two members of the category of squishies – octopuses and cuttlefishes – lay their eggs close to shore, while squids lay their eggs out at sea.[5] Such proximity to the solid land explains why the cuttlefish has a blacker ink than the squid, since the darker version carries more earthy elements. Their proximity to shore is also reflected in the fact that octopuses have 'twining feet' to protect them from predators, and

Modern photo of squid eggs – less artful but more accurate.

occasionally to walk on land (a detail repeated in later histories). Unlike either of these, only squids live out at sea, finding protection in the deeps.[6] With their slighter connection to the land, squids have less earthy substance and less need of protective devices, and are more uncertain to Aristotle.

For the most part, Aristotle refers generically to the squishies (*malakoi*), and only occasionally distinguishes the octopus (*polúpous*) or the cuttlefish (*sepía*) by name. When he refers to squids, he offers another division. On the one hand he says that 'both the *teuthos* and the cuttlefish are short-lived: only in a few instances do they reach their second year', and then just a few lines later on he says, 'The male *teuthis* differs from the female: the latter, as can be seen if one pulls open the hairy region and looks inside, has two red objects resembling breasts; these are not evident in the male.'[7] The standard Greek–English lexicon translates *teuthis* as 'calamary or squid, *Loligo vulgaris*' (and also notes that the same word might refer to 'some sort of pastry'), and for *teuthos* offers this: 'calamary or squid, of a larger kind than the *teuthis*, perhaps *Todarodes sagittatus*'.[8] While the first term appears in other, non-Aristotelian sources, such as Aristophanes and Galen, the latter term appears to be confined to Aristotle alone. When commenting on the squishy, watery quality of cephalopods, he observes that, in addition to a strong structure in their flesh, two groups possess something almost akin to the bones of fishes. In the cuttlefish is found the 'pounce' (*os sepiae*), and through squids runs the much thinner and softer 'pen' (*gladius*).[9] The Greek word translated as 'pounce' (which can refer to the ground-up cuttlebone that is used to prevent ink – coming from the same animal – from blotting) is *sepion*, which seems straightforward enough. The 'pen' or '*gladius*' refers to the cartilaginous structure in squids that was, in fact, used as pens, and for which the Greek word is *xiphos*, translated as 'sword-shaped bone, as in the cuttlefish

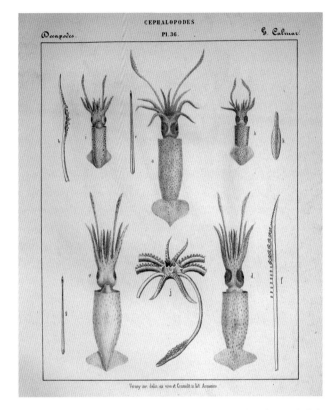

An 1851 illustration of *Loligo coindetii* (now called *Ilex coindetii*), the southern, or broadtail shortfin squid, with anatomical details.

(*teuthis*)'. For the Greek–English lexicon, a squid and a cuttlefish are the same animal.

Of course, confusions can abound whenever modern translators struggle to organize ancient references, and 'bone' certainly calls to mind the cuttle long before the much squishier squid. But confusion between the cuttlefish and squid (the real *teuthis*) actually begins much earlier than in modern translations. Most of the natural historians who relied on Aristotle mix up his distinction of the cuttlefish (*sepía*) and squid, and overlook the further division of squids. As for the two sorts of squid, Aristotle distinguishes the *teuthos* from the *teuthis* by its size, stating that 'specimens are found as long as five cubits.'[10] Setting the Greek cubit at roughly 45 cm (18 in.) would give us a *teuthos* of around 2.25 m (7½ ft) in length, a measurement that has induced modern fans of the giant squid to find their earliest reference in Aristotle. But, though he did travel widely, Aristotle almost certainly did not see the Atlantic, and giant squids have not been spotted in the eastern Mediterranean – at least in modern times.[11] The most certifiable distinction would be to say that *teuthis* refers to inshore squids, and *teuthos* to the larger deep-water squids, such as the *Todarodes sagittatus*, which often reaches lengths of a little over 60 cm (2 ft) and is common throughout the Mediterranean. These two categories of squids would include numerous genera and species that Aristotle did not distinguish. His smaller inshore squids would probably have included those still prominent in

Comingio Merculiano's 1896 depiction of the European arrow, or flying squid (*Todarodes sagittatus*).

fish markets, and the great natural historian could also have seen them swimming in the shallow waters close to shore. The larger squids would have been less common, and Aristotle almost certainly would not have seen them in their natural habitat of the deeper offshore waters. While he might have encountered some specimens in the catches brought to shore, he would have mostly relied – like the naturalists and biologists ever since – on descriptions given by the fishermen who caught them or merely glimpsed them.

Well into the twentieth century, this terminological fluidity persisted among the common names of the squishies, creating a long tradition of uncertainty over which particular creature a natural historian might mean. This fluidity led more than one modern commentator to claim classical writers refer to cephalopod species – particularly the giant squid – that probably did not inhabit the ancient Mediterranean. The same fluidity has led people throughout the millennia to refer to one group when they seem to mean another. With the best of intentions, natural historians, as well as poets and novelists, use almost all available cephalopod names interchangeably.

D'Arcy Wentworth Thompson, one of the foremost commentators on Aristotle in the early twentieth century, attempts to clarify the terms by emphasizing the size difference between *teuthos* and *teuthis*. Whereas for Homer 'poulypous is apparently so far generic as to include the cuttlefish and the calamary', Thompson argues that Aristotle was not blind to the distinctions among species.[12] He lists some common names for squids in Mediterranean languages to give the difference a solid grounding. The smaller species he translates into later languages as *preke* (Old English), *calmar* (French), *calamaro* and *calamaii* (Italian), and the *teuthos* as 'a larger kind . . . the *totaro* of the Neapolitan fishermen; in Modern Greek *trapsalo*'.[13] In providing all these terms from different languages Thompson follows the same effort by natural historians going back several hundred years. The aim is to foreground regional familiarity with squids, even alongside the admission to being confused by the apparently conflicting terms. One regional name, *preke*, does, however, seem particularly descriptive when we find that the sixteenth-century French naturalist Guillaume Rondelet records fishermen referring to a certain sea creature as '*vit-volant* [flying-prick]; when alive it swells up and makes itself thicker; being deprived of life, it becomes totally withered and limp'.[14] Of all the names applied throughout history, this is one of the few that cannot be easily applied to cuttlefishes or octopuses.

Perhaps what stands out most in Aristotle's accounts is that they should distinguish so few genera of squids – or octopuses and cuttlefishes, for that matter – since numerous species of all three cephalopods abound in the Aegean. One reason is the ancient Greeks' aversion to the wine-dark sea. The highly entertaining classicist Norman Douglas observes that 'the ancients . . . had a knack for keeping the sea, as such, in its proper place.' From his study of the ancient collection of texts known as *The Greek Anthology*, Douglas comments that the poets represented therein 'had little of our sense of the wonder or beauty of this useless and monotonous mass of water . . . The average of all the epithets applied to it may be summed up in the word *perfidious*.'[15] While octopuses and cuttlefishes might be observed close to shore, squids, as Aristotle points out repeatedly, live farther out at sea. Even the inshore species suggests to him an animal from the perfidious realm. D'Arcy Wentworth Thompson was the first reader to make the important deduction that Aristotle spent two years in Lesbos, where he examined many of the marine creatures described throughout his natural history. Thompson says, 'it is to the lagoon at Pyrrha that Aristotle oftenest alludes,' and that 'here among the cuttle fishes was found no octopus, either of the common or of the musky kind.'[16] Thompson might have gone on to point out that the lagoon (or Gulf of Kalloni as it is commonly known) would have contained only the smaller inshore squids, but none of the larger sort. What Aristotle actually was able to observe from the shore, or as he kept his feet dry on the ships carrying him between Athens and Lesbos, was probably limited to what would have been provided by the fishermen telling him of the beasts they had encountered, or heard of, from the dark and perfidious sea.

My own speculation about how far Aristotle ventured onto or into the deeper waters accounts for the descriptions that have the quality of legend, of seamen's tales. In his account of squishies' feet, he asserts that 'all of them have eight'; and then he qualifies this claim by pointing out a 'peculiarity' in cuttlefishes and the two species of squids who all have an additional two long tentacles, used to convey food to their mouth, 'and when the weather is stormy they throw them out to fasten on a rock like anchors

Giacomo Franco's 1597 Venetian map of 'Metileme' [Mytilene], now known as Lesbos.

and ride the storm'.[17] A basis for this claim of tentacular anchorage, apart from Homer, remains unstated, of course, but finds a telling echo in the comment by Aristotle's successor at the Lyceum, Theophrastus, that a squid serves as a sign for a coming storm. Since that particular warning stands alongside numerous other meteorological signs – such as asses shaking their ears, or cattle lying on their right side – Theophrastus appears to have expected frequent storms. But it also suggests that squids showed up rarely enough (at least in the surface waters and during the day) that the appearance of one had to carry some kind of significance, and that portended something unsettling. Such portentousness continues into modernity, and is a constant in fishermen's tales.

Historian Nicholas Purcell says that to the inhabitants of the ancient Mediterranean, the sea was 'ugly and lonely, a desert in which the human is wholly out of place'.[18] Other commentators have pointed to 'the sea's own power to swallow and conceal a man completely', citing the common poetic epithet of the 'gullet of the salt sea'.[19] And since, of all the animals eaten in the ancient Mediterranean, sea creatures were the only ones that ate people, they gave rise to further anxieties. Such a frightening possibility is depicted on a fragment of ceramic from the eighth century BCE showing drowning sailors being eaten by varieties of fish, underscoring the dangers of sailing and of falling into the salty gullet swarming with more gullets.

Eating fish, and particularly squid, it seems, could connote a moral ambiguity, giving rise to various prohibitions on religious grounds (for an adherent to the doctrine of transubstantiation, eating an animal that eats humans raised the alarm of cannibalism). Modern Greek restaurateurs will remove fish from their menus for a day whenever a drowned corpse is discovered nearby.[20] Aristophanes has one of his characters in the comedy *The Knights* level a pertinent curse at the politician Kleon:

May your skillet of squid be standing ready and sizzling; and may you be about to propose a motion concerning the Milesians, and make a talent if you carry it; and may you therefore make haste to fill yourself with the squid and still get to the Assembly in time and then, before you have eaten it, may a man come for you, and may you in your eagerness to get the talent choke on the squid as you eat it.[21]

The comically elaborate imprecation builds on the assumption that the politician is corrupt, and his corruption is made evident in his hunger for both bribes and cephalopods. Eating squids constitutes an ancient eating disorder, of consuming what should not be eaten, what is not considered food. Because fishing in the ancient sea was both dangerous and uncertain of success, an appetite for squids signified the power to command men to risk being swallowed, and the decadence of defying social taboos with impunity. Nonetheless, the first-century CE cookbook *De re coquinaria*, ascribed to the Hellenic cook Apicius, includes the advice to 'put the stuffed cuttlefish in the boiling pot so that the stuffing can come together.'[22] The stuffing would probably have consisted of brains from a land animal, perhaps a sheep or more exotic mammal, and its reputation for being both indigestible

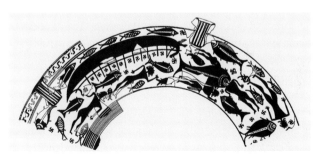

Design from an 8th-century BCE *krater* vase of shipwrecked sailors eaten by fish.

Squid cooked in its own ink.

and an aphrodisiac rested on the combination of terrestrial organs and the sea-borne *preke* that might have digested men's organs.[23] At the same time, the recipe suggests that the attitudes of later Greeks had begun to change.

Decapods – or much of any seafood – did not commonly appear on the Greek plate before the fifth century BCE. In fact, only four images of fishmongers are known to exist in Greek vase painting.[24] Athenaeus, the third-century BCE author of *Philosophers at Dinner*, quotes the fourth-century BCE gourmand Archestratus with a barely repressed sneer: 'Archestratus who travelled round all lands and seas to satisfy his gluttony says, "Squids are to be found in Dium of Pieria by the surge of Baphyra. And in Ambracia you will see many."'[25] Since Athenaeus' text is the only source for the fragments of Archestratus, the derisive comment shows the view Hellenistic moralists took of people who went to such lengths to procure culinary delicacies. The writer on extravagant cuisine would certainly be criticized for knowing the choice location for squids, since the consumption of squishies denoted an indulgence condemned by the prudish creed of Plato to which Athenaeus adhered. Living in Syracuse, Archestratus had freed himself from the older Greek aversion to the wine-dark sea, as his fragments indicate a comfortable familiarity with it, which might explain his similar familiarity with squids and their location: Dium can be found in southern Macedonia, which would have required a sea journey around the Peloponnesus. This area would be close to the 'Sepiad headland', where, according to Herodotus, the Persian king Xerxes anchored his fleet on the way to invading Athens; he paid due obeisance to Thetis, mother of Achilleus, because one of the shapes she assumed was that of a cuttlefish, when Pelops was able to seize her at this very spot.[26] Thetis, daughter of the sea god Nereus, represents the changing, liquid quality of the ocean. This aspect, along with its darkness, led even the later Greeks to consider dangerous, man-swallowing water as the most primordial of the elements, because it has yet to become divided and organized into solid forms.

Elena Recco, *Still-life with Fish and Sea Animals*, 1695–1705, oil on canvas. These marine creatures would delight the most decadent Greek gourmand.

This point is worth pausing over, for I believe it reflects in multiple ways on ancient accounts of squids. If these animals come from a realm that most people would turn away from, and if they had held a moral (or at least dietary and meteorological) stigma, then Aristotle's interest in them would reflect either a personal or cultural shift even before Apicius. In the fourth century BCE, when Aristotle lived, Athens ruled the eastern Mediterranean as its famed 'thalassocracy', using its navy to dominate the waters. Athenian sea power would certainly have enabled Aristotle – and Archestratus – to travel over the sea, and would have also encouraged fishermen to venture farther from shore (though not terribly far, as the classical navies fought their battles within sight of land). But political domination does not extend below the sea's surface, and the deeper seas could still arouse fears in the most intrepid philosopher. Aristotle based his philosophy on direct experience, and thus could offer detailed descriptions of the creatures he encountered, such as the cuttlefishes and inshore squids (those he called *teuthis*) he would have seen in the Gulf of Kalloni and dissected. The larger squids, those termed

Rondelet's calamary, 2nd century CE, distinct from the ecclesiastical monsters only in descriptive terms.

teuthos, swam beyond the reach of his direct experience, and so his description and analysis of them is necessarily uncertain, allowing his successors to overlook the distinction and even merge them with cuttlefishes. His willingness to describe squishies, along with the Athenian desire to rule the waves, surely helped dispel the moral aversion to (though not the fear of) squids and the deep sea.

The Romans were free of any compunction about entering strange territories or about consuming whatever they found there. Pliny's first-century CE accounts of marine life follow Aristotle's in their reliance on seamen's tales and cultural whims, however. The category Pliny devises, 'bloodless fishes', adheres to the Aristotelian view that cephalopods flowed with something between human blood and divine ichor. Of the three sorts of bloodless fishes, the first variety Pliny calls 'soft' (*mollia*), which includes, not surprisingly, the 'cuttle-fish', or *lolligo*, the sepia, or *saepia*, and the polyp, or *polypus*. And so we see the lexicological blurring sustained. The Latin name, *lolligo*, has become an official scientific classification for a large genus of squids, not cuttlefish. Pliny's English translator uses the terms 'cuttle-fish' and 'sepia' to refer interchangeably to *either* a squid or a cuttlefish. So, the modern reader who dozed through Latin class must keep a sharp eye. Still, a blurring does happen. When Pliny says that the 'two varieties of cuttle-fish' copulate 'with their tongues' he seems to have both cuttles and squids in mind, since he elsewhere drops Aristotle's teuthic division. And when he says, 'the sepia does not occur in the Black Sea', but that 'the cuttle-fish is found there', the reader may feel that the confusion does not lie entirely with the translator, for no species of modern squid appear in the Black Sea.[27]

A reader of classical literature on squids could almost fall into despair over trying to untangle this terminological confusion between squids and their distant relatives. The terms here do

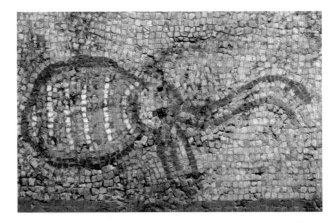

Mosaic of a cuttlefish (*sepia*) from a Roman villa at La Pineda, Catalunya. Pliny confused squids and cuttlefishes.

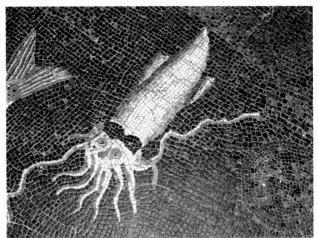

Roman mosaic of a squid among its fellow sea creatures, 1st century BCE.

remain consistent, in that '*lolligo*' gets Anglicized as 'cuttle-fish' throughout, so that the translator – unaware perhaps that *lolligo* has a decided place among modern squid categories – chooses the name more familiar to him. Still, this belief that a cuttlefish is a squid, while a *sepia* remains apart, indicates a cultural slippage, one that I care less to correct than to appreciate as a notable aspect of the squishy tradition we have inherited, for it has consequences.

The interchangeability of terms might not reflect confusion but rather the classical perception of fluidity that still influences us. When Aristotle stresses that squishies hold the aqueous qualities of their environment, he establishes a philosophical principle that had already existed as common sense. And so Pliny, in observing the large size of many sea creatures, explains that 'the obvious cause of this is the lavish nature of liquid . . . Because the element receives generative causes from above and is always producing offspring, a great many monstrosities are found.'[28] To Pliny's mind, the nature of liquid, 'lavish' in its nourishment and influence, causes the creatures within it to multiply in number as well as in shape and quality, blending and flowing together and in various directions. The forms of land creatures meet their limits through the solidity of their components and their environment, while the sea creatures fluently expand into larger and stranger forms unseen in other realms (and unseeable in general, so long as they remain under water). Pliny goes on to forge an analogy that

The arrow or flying squid (*Todarodes sagittatus*) from Férussac and d'Orbigny's *Histoire naturelle, générale et particulière des céphalopodes* (1835–48).

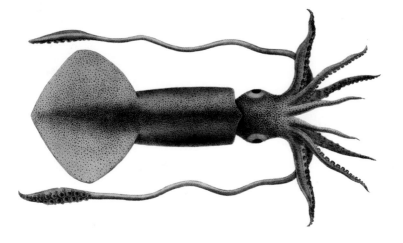

reveals his speculations about what lives in the deep darkness. He says, 'everything born in any department of nature exists also in the sea, as well as a number of things never found elsewhere.' Assuming the winey sea contains creatures he has not seen, he can imagine them only by projecting marine versions of terrestrial animals. And he adds a few more just to emphasize the strangeness of life under water.

The analogy raises problems, however, when Pliny asserts that 'the cuttle-fish [*lolligo*] even flies, raising itself out of the water, as also do the small scallops, like an arrow.'[29] This characteristic strikes such an odd note among classicists and natural historians alike that even well into the twentieth century, Thompson includes a reference to *teuthis* under the entry for *chelidon*, or flying fish. This learned scholar says, 'the mention of *teuthis* along with flying fishes by Aelian and Oppian is puzzling, especially as it is repeated by Pliny . . . for a cuttle-fish has seldom been said, or seen, to "fly"'.[30] Thompson's confusion of cuttlefishes and squids culminates in a puzzlement that would not have concerned the Roman observers of marine life, or even modern squid-watchers. In fact, a common name for *Todarodes sagittatus*, which he identifies with Aristotle's *teuthos*, is the European flying squid.

The poet Oppian, who probably wrote his treatise on fishing, *Halieutica*, in the second century CE, relied mostly on Pliny for his information, and he describes the marine creatures primarily with an eye on how they might be caught and eaten. His reference

Illustration of *Ommastrephes sagittatus* (now *Todarodes sagittatus*), the arrow or flying squid, whose common name troubled unbelieving naturalists, 1870.

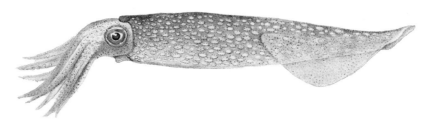

to the flying *teuthis* combines hints about cuisine and couture in the anonymous seventeenth-century translation of his text as 'Lolligo, the Sleve, a flying fish'. At the time of this translation, a 'sleve' was not necessarily sewn onto a fuller garment, but often consisted of a separate tubular covering for the arm. In the wardrobe of one's lexicographical mind, a squid could be described as ten arms radiating from such a 'sleve'. Like the flying '*preke*', the 'sleve' looks squishily empty when brought to shore, begging to be re-stuffed, as Apicius recommends. Even today, cooking expert Alan Davidson says, 'the squid might well have been created expressly for cooks to stuff, so conveniently is its body shaped for this purpose.'[31] Davidson and Apicius fill the sleeve in their haute cuisine, while others dress it differently.

Guillaume Rondelet's sixteenth-century *Libri de piscibus marinus* largely repeats Aristotle's categories, but this is where the tradition of natural histories of squids takes a notable turn. The Greek and Roman writers described what they saw with a familiarity that came from direct encounters as well as from references made by fishermen (and, in the case of Oppian, from fishing for the 'sleves' he intended to eat). The natural historians of early modern Europe serve two aims in their accounts. Like their classical predecessors, they want to identify what a squid is, and particularly in their own culture that has invested them with its distinct values. At the same time, they need to establish their authority as natural historians by demonstrating a knowledge of Aristotle and Pliny. Just as D'Arcy Thompson confuses himself by referring to a cuttlefish when he means a squid, the early European naturalists update the classical accounts in the terms of local usage. The result is a case study of the transformation of creatures people had actually encountered (and eaten) into mythic monsters. This transformation of the ordinary into the bizarre – this monstrification – takes place only because of the slipperiness

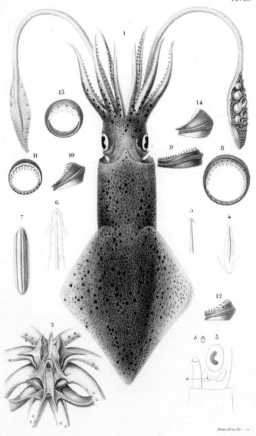

PL. 18.

1-12. Loligo vulgaris, *Lam.*

13-14. L. —— Gahi, *d'Orb.*

Imp. de J.Delarue.

of descriptive terminology. In it we can see a genuine effort to catalogue a cultural understanding of an animal that would challenge most people's descriptive skill.

Rondelet describes two species of loliginids, *Loligine parua* and *Loligine magna*, along with a *Sepia* and *Sepiola*. The two loliginids repeat Aristotle's division of *teuthis* and *teuthos*, attending almost entirely to the size difference. But the reason Rondelet's history has garnered attention is that prior to these entries, it describes two bizarre squid-like creatures. The first is the *pisce monachi habitu*, illustrated by a humanoid form in a monk's habit, bearing upwardly curled spiny appendages instead of human arms and hands, and what looks like a full-length spiny skirt also ending in upward curls. Following right after this 'fish in a monk's habit' comes the *pisce episcopi habitu*, which has a less human appearance, but is still not immediately recognizable as a modern (or Aristotelian) squid. The hybrid forms of these two monsters reflect a continued reliance on fishermen for information. That is, sailors recount seeing creatures emerge from the water looking like a habited monk or a mitred bishop, and the drawings follow. Rondelet, like natural historians before him and like marine biologists after him, has given literal weight to descriptive similes that might not sound so bizarre when spoken, such as, 'the creature was tubular, like a bishop's mitre', or 'its skin made me think

foramē in fronte in aſperam arteriam deſinens. Item cum branchia-
rum ſciſſura, cum pedibus, vnguibus, ſquamis quibus omnibus certiſ-
ſimū eſt balænas carere. Pręterea ex teſtudine marina finxerūt mon
ſtrum à quo vitulus marinus auriculas habens præter veritatē, deuo-
rátur. Nec minùs abſurdè orcā pinxerūt, quæ balænā perſequitur, &
naues euertit. Abſurdiſſimè verò phyſeterē forma equi cum patulis
naribus & fiſtulis duabus prominentibus, cum auriculis aſininis, cum
lingua longiſſima & prominente. Nec minùs monſtroſam finxerunt
Scolopendrā cetaceam quadrato capite, promiſſa barba. Nō minore
errore porcum marinū repræſentarūt, & orcę ſpeciem quandam cui
à celeritate nomen poſuerunt, & vaccam marinā, & monſtrum aliud
rhinoceroti ſimile. Hæc ſibi in chorographia illa Septentrionali per-
miſerunt pictores, quę tamen vera nonnulla continet, vt aſtarum ma
rinum, onocrotalum ſatis rectè expreſſum, & terreſtria aliquot.

De Piſce monachi habitu.

C A P V T XX.

 NTER Marina mōſtra referemus & illud quod no-
ſtra ætate in Nortuegia captum eſt mari procelloſo, id
quotquot viderūt, ſtatim monachi nōme impoſuerūt:
humana facie eſſe videbatur, ſed ruſtica & agreſti, capite raſo & lęui, humeros cōtegebat veluti monachorū
noſt

De Pisce Episcopi habitu.

CAPVT XXI.

ONSTRVM Aliud multò superiore mirabilius subiungo, quod accepi à Gisberto Germano medico, cuius antè aliquoties memini, quod ipse ab Amstero-damo cum literis acceperat, quibus ille affirmabat an-no 1531. in Polonia visum id monstrum marinum Epi-scopi habitu, & ad Poloniæ regem delatum, cui signis quibusdam si-gnificare videbatur vehementer se cupere ad mare reuerti, quò dedu-ctus statim in id se coniecit. Sciens omitto plura, quæ de hoc monstro mihi narrata sunt, quia fabulosa esse arbitror. Ea est enim hominum vanitas, vt rei per se satis mirabili præter verum plura etiam affingát: ego qualem monstri Erconem accepi, talem omnino exhibeo. Vera ea sit an non, nec affirmo, nec refello.

The *pisces episcopi* or 'sea bishop', Guillaume Rondelet's illustration based on fishermen's descriptions, 1554.

of a monk's habit' (which could be said to drape loosely over the body like a sleeve).

In the eighteenth century, the Scandinavian clergyman Erich Pontoppidan published *The Natural History of Norway* (1755). From the start of his descriptions of squids, he admits to having diffi-culty coming up with a coherent set of details, and even fumbles with the name, listing it as 'the Spoite, Bleksprutte, the Sepia, or Ink-fish, called also by some the Sea-gnat', and finally mentioning that 'some authors call it by the name of Sepia, or Loligo'. Looking

Rondelet's calamary, distinct from the ecclesiastical monsters only in descriptive terms.

Der dreyzåhendt teil von

Loligo magna, Lolium, Teuthon. Ein grosser schmaler blackfisch/Ein Meerschreyb zeüg/ Meer lülch/ grosser horn kuttel/ måsser kuttel/ såder kuttel.

Von irer gestalt vnd mancherley grösse.

Diser kuttel fisch ist de vorbeschribnen brei-ten black fisch gantz ånlich/in aller seiner gestalt/allein daß er etwas lenger/schmeler oder raner ist/ zů end spitzig. Sein bein oder schwårdt so er hat ist dünn/schmal/durchscheyned/seit råchtes horn ist dicker dann das linge/ spritzt auch hår-auß/schwartze farb/võ wel-cher er black fisch genennt wirdt. Söllend an ettlichen orten mit der lenge kommen auff 5. ellen. Das weyble wirdt võ dem månle erkent/so das weyble auffgeschnit-ten/so werdend zwen zing͂/derm oder gmåcht gesåhen/ tåglich zů der frucht/welche die månle manglend. Sy habend auch etwas purpur farbes gesaffts in jnen/ auff welcher vr-sach sy gekocht/rotlåcht gesåhen werdend.

Von der anderen gestalt.

Loliginis maioris forma Venetijs efficta.

Ises ist gantz ein schöne/grundtliche ge-stalt/des grossen schmale Blackfischs zů Venedig abconterfetet.

Von

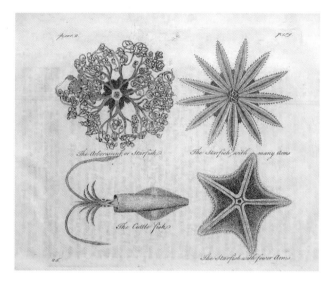

Erich Pontoppidan's 'cuttle fish', from his *Natural History of Norway* (1755).

to his classical authorities, he makes the telling (and maybe even traditional) admission that he cannot understand Pliny's distinction between the sepia and loligo. Adding to this difficulty, he says that squids are virtually incomprehensible to those who have never seen one, because they continually shift their appearance by moving their arms and changing colours. All this confusion and uncertainty makes Pontoppidan admirably meticulous in admitting that creatures with ten legs radiating from their heads seem all but inconceivable. He tells us a great deal about the source of his information when he says that 'in the last century our peasants looked upon the Cuttle fish to be a dangerous and ominous creature: they called it an amazing sea-prodigy, when they catched one near Katvig in Holland, in the year 1661.'[32] This 'cuttle', 'spoite' or 'Loligo' appeared ominous to the men who 'catched' it because they could find no other way to describe the shifting appearance of an animal that changes colours. The multiplicity of local names

Pontoppidan lists underscores his own confusion over Pliny's distinction: the phrase 'amazing sea prodigy' expresses the feelings of uncertainty over the kind of being the peasants, as well as the learned Pontoppidan, found themselves encountering.

Pontoppidan himself is very much a man of the Enlightenment, and thus is insatiably curious about the world even while anxious about the overwhelming discoveries showing that world to be far from stable. As he closes his account of the Scandinavian waters, he says that 'not only the incomprehensible numbers, but the variety also much exceeds, by what we can judge the species of land animals.' His modern faith that reason can organize the entirety of animal varieties into stable categories appears plainly shaken by the confrontation with the inconceivable Loligo. 'On the one side', he says,

> we ought not to be too credulous and believe the idle tales and improbable stories that every Fisherman or sailor relates . . . yet I am of the opinion, that the other extreme deviates as far from the truth, namely when we will not believe things strange and uncommon, tho', according to the unchangeable law of nature, possible, because we cannot have so evident and clear a demonstration of it as we might.

This generosity mixed with scepticism is what makes Pontoppidan, like so many children of the Enlightenment, able to look to fables, myths and 'improbable stories' as the bearers of some sort of hidden cultural truth – if not actually empirical or useful, then at least of a kind that challenges the old parochial inhibitions on what is to be thought possible, or which inhabitants of our world might be spoken about. 'One might as well doubt whether there are Hottentots,' he says.[33]

The great Sea Serpent, according to different Descriptions

Pontoppidan's Great Sea Serpent, described alongside the 'cuttles'.

Pontoppidan also hesitates to describe the 'Kraken', listed among the 'Norwegian Sea-Monsters', which he defines simply and in accord with the monstrous tradition as 'those Animals of enormous size and uncommon form, which are sometimes seen in the ocean'. As with the less gigantic squids, he seeks to avoid the imputation of being overly credulous about the existence of enormous animals; but he also, as a clergyman, insists that even monsters can be taken to illustrate the 'great Creator's wisdom, power, and glorious œconomy'.[34] As insatiably curious as Aristotle, Pontoppidan also knew of Pliny's account of the sea's lavish generative forces that create monstrosities found nowhere else. As a scholar therefore, he is obliged to listen dispassionately to fisherman's tales and to align those with his readings in authoritative texts. The sea monsters reflect his effort to reconcile Aristotle's obscure deep-water *teuthos* with equally tantalizing descriptions offered by unlettered mariners of the colour-shifting animals they had seen in the stormy seas. Pontoppidan can find room for the improbable stories, but he worries over how they upset Aristotelian taxonomy, which seems unable to accommodate such a variable creature. His only recourse is to gesture to the vaster wisdom of the Creator, whose 'glorious œconomy' is not always plain to mortal natural historians.

Following the hesitant and enlightened Pontoppidan came Pierre Dénys de Montfort, the young French noble who attached himself to the Revolution and changed his name to Dénys-Montfort.[35] Hoping to extend the Revolution's rational re-assessment of nature and *le peuple* into a career in natural history, Dénys-Montfort interviewed a community of American whalers who had taken up residence in France. His aim was less to preserve a cultural tradition than to uncover a marine analogue of the revolutionary forces that had titillated and frightened Europe. He recounted the whalers finding 'the limb of an enormous (quid) octopus' protruding from the mouth of a dying whale; the limb, 'accurately measured with a fishing line, turned out to be seven fathoms, or thirty-five feet [10.5 m], long'.[36] The parenthetical 'quid' is Dénys-Montfort's emendation, and indicates an attempt to correct the whalers' belief that the arm had come from a giant 'calamary'. The persistent ambiguity, or squishiness, of cephalopod terminology had apparently got the better of the citizen naturalist, as he sought to level the taxonomic differences into a general council of cephalopods seizing control of the watery kingdom. The majority of modern naturalists deride Dénys-Montfort as a sensationalist who believed in the mythical Kraken and the giant octopus, or poulpe.

Most notably, or perhaps most notoriously, Dénys-Montfort included an illustration of a ship under attack by the colossal poulpe. This illustration was based on a painting donated to the chapel of St Thomas in the village of St Malo on the northern coast of Brittany by a group of sailors who had survived such an attack by a kraken. Later in the same century, English naturalist Henry Lee harrumphs, 'in a drawing fitter to decorate the outside of a showman's caravan at a fair than seriously to illustrate a work on natural history, he depicted this tremendous cuttle as throwing its arms over a three masted vessel, snapping off its masts, tearing down the yards, and on the point of dragging it to the bottom.'[37] Never mind Lee's anti-republican bias, the painting, as well as the chapel of St Thomas, appear never to have existed in St Malo.[38] Publishing this one illustration, as though it provided empirical affirmation, is the reason Dénys-Montfort fell into such disrepute, for otherwise he could simply lay claim to repeating the legends that had long circulated about a giant cuttlefish, or poulpe, or kraken, or sea-gnat. But just as natural historians before him extended the biases and fears of their culture, Dénys-Montfort made claims that reflected the sensational aspirations of a revolutionary era in which people could express ambitions previously unthinkable, and discoveries previously deemed fantastical. And so naturalists even outside of Scandinavia began to investigate accounts of the Kraken, mermen and sea serpents, as though they might exist outside of legend and folklore.

Fourteen years after Dénys-Montfort's enthusiastic account, the Scottish natural historian James Wilson published a two-part article in *Blackwood's Edinburgh Review*, 'Remarks of the Histories of the Kraken and Great Sea Serpent'.[39] Of the 'two most famous monsters described in history', namely those whose depiction belongs on a showman's caravan, Wilson says that 'although the existence of the latter has been proved by the most satisfactory evidence, within a very recent period, the former is still regarded as a mere chimera.' From this, Wilson goes on to cite authoritative accounts of the Kraken, linking them to the 'great Polypus, or Cuttle Fish'. The 'most zealous' of his sources is Dénys-Montfort, whom Wilson praises for providing numerous eyewitness accounts of the Kraken's appearance around the world.[40] What matters to Wilson is less the outlandish and dubious illustration than the

The *Poulpe colossal* by Dénys-Montfort, 1802. This image won him the disdain of his English successors, and its supposed original never existed.

Pl. XXVI.   *T. 2. P. 256.*

Denys-Montfort del.   *E. Voysard S.*

LE POULPE COLOSSAL.

genuine recording of what the American whalers had said. In the second part of his article, Wilson recounts the natural history of sea serpents, enumerating many of the same details that would lead writers like Lee to connect them to giant calamaries at the end of the century. To the modern eye, sea serpents, mermaids, giant polyps, sea monks and the Kraken all bear enough resemblance – and perhaps outlandishness – to be grouped in a single squishy taxon: that of myth.

The significance of Wilson's article lies in the effort of a serious natural historian to determine whether or not the most prominent legends of the sea had any merit as fact. Like his predecessors going back to Pliny and Aristotle, Wilson can only recount those legends, and lend them a credence based on generosity more than verifiable documentary evidence. Through the nineteenth century, the legends refused to give way to the sniffing and pooh-poohing of the new empirically grounded scientists. To the trained rational mind, what seemed most unlikely of all was that a giant version of the relatively ungigantic squid could exist; whether a large cephalopod could wrap its tentacles around a ship became secondary to the more pressing question of the Kraken's existence.

There is also the fact that first-hand sightings of krakens in the sea and stranded on Icelandic and Scottish beaches became increasingly common by natural historians who no longer had to rely solely on the accounts of sailors. By mid-century the Danish natural historian Japetus Steenstrup had become sufficiently attuned to these repeated sightings of a creature that was supposed to reside only in the untrained minds of inarticulate seamen that when he received the large beak of a beached squid from Jutland, he understood its significance and gave the giant a place in the official taxonomy by naming it *Architeuthis monachus*, referring to Rondelet's 'fish in a monk's habit'. On 26 November 1854 Steenstrup delivered a lecture to the Danish Natural History Society in

Japetus Steenstrup (1813–1897), the man who brought the legends of giant squids into scientific reality.

which he first proposed the official name of the giant squid. He opened the lecture with a comment echoing the caution of three centuries of natural historians: 'mermen and mermaids are considered by many to belong to the kind of creatures with which the naturalist should not deal in order that he should not endanger his good name and reputation.' The lecture focused primarily on the 'sea monk' described by Rondelet and others that had washed ashore in 1550 and seemed similar to the one beached in Jutland. Steenstrup worked through an exegesis of the different accounts that describe the creature sleeved in the habit of a monk, and with a shaved head. One account recorded, Steenstrup pointed out, that the king ordered 'this abominable creature' to be buried in order not to 'provide a fertile subject for offensive talk'. Even without knowing Theophrastus, His Highness understood that such abominations could warn of rough seas and revolutions. Based on the details in all the accounts, Steenstrup concluded that they described the same event, and that the sea monk was a non-signifying cephalopod. He provided visual support by aligning the two sixteenth-century illustrations of the sea monk with an analogous drawing that would allow for a squid's fins and arms, with the two tentacles lying underneath. The bald head he connected to the smoothness of the squid's sleeve-like mantle, and the scaly monk's habit to the presence of colourful chromatophores. When he posed the rhetorical question 'But is our Sea Monk now a common squid?!', he made a significant turn in the lecture (and in natural history), focusing on the size documented in the sixteenth-century texts as 'four ells', which can be modernized as roughly 2.5 m (8½ ft). Again he admitted that most naturalists would consider accounts of giant cephalopods with no less suspicion than the sea monk. He then reminded his audience that two such giants were found on the coast of Iceland after storms in 1639 and 1790. He drove the point home by dramatically displaying a specimen jar of the beak he had received from Jutland, drawing attention to its size. With that flourish, he announced that the monk should be recognized once and for all as a squid, and one as large as those described in legends.[41]

Steenstrup's presentation had a decisive effect, and was bolstered by the beachings of several more giant squids in subsequent years. In the most prominent sense, Steenstrup ended the long tradition of natural historians reporting on a legendary creature (or set of creatures) as an uncertain member of the zoological population and a figure in the regional folklore. By displaying the beak of the Jutland squid, he was able to bring the giant species under the scrutiny of modern empirical science, wresting it from the uncertainty of natural history that had relied on second-hand

Sømunken efter Rondelet.   Loligo - Blækſprutte fra Kattegattet.   Sømunken efter Belon.

The images
Steenstrup used to
prove that the
sea-monsters were
actually squids.

accounts. Even though marine biologists would not observe the
giant squid in its habitat for another 150 years (and even though
modern biologists believe that *A. dux* probably does not willingly
swim into the shallow North Sea[42]), the beak provided a firm basis
for the giant squid to enter the official taxonomy as *Architeuthis
dux* – oddly enough, not *Architeuthos*.

In a slightly less prominent sense, one that is no less conse-
quential, however, Steenstrup brought squids generally into a
kind of terminological regularity. Previously the terminological
slippage had allowed for a kraken to be called a sea monk or a
devil fish, or poulpe, or polypus, or colossal sepia, or octopus, for
all these terms referred in a general way to the monsters described
in legends. After the terminological assignation of *Architeuthis*
(whether of *monachus* or *dux*), all but one term became incorrect,
even if only because they came from regional legends. While
debates continued through the nineteenth century over whether
the legendary Kraken and sea serpent were a mere quotidian

calamary, the gullibility attached to an acceptance of folk tales changed into a rational acceptance of diverse life forms, so long as they could be catalogued through commonly accepted descriptive terms that refer to a limited set of squid characteristics. No less rare than the Kraken, *Architeuthis* now belongs in the taxonomy of level-headed scientists rather than in the wild-eyed histories of enthusiasts like Dénys-Montfort. As natural history, blending local descriptions with the tradition of classical learning, gave way to the scientific discipline of marine biology, the confusions over *teuthis* and *teuthos*, and squid versus cuttlefish, disappeared behind characteristically modern concerns over verifiable data garnered through countless dissections and the commercial value of various squid species. All that makes for a different story, disclosing the assumptions of another culture and era – our own.

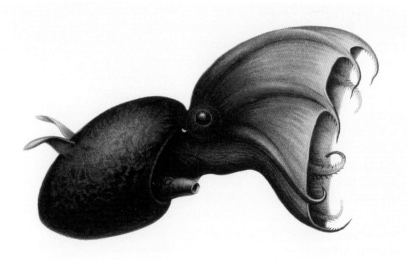

2 Modern Teuthology

In 2012 marine biologist Tsunemi Kubodera led a deep-sea expedition off the Ogasawara Islands in the western Pacific to film the giant squid (*Architeuthis dux*) in its natural habitat. Kubodera and his crew became the first humans to see the elusive giant alive and swimming in its customary depth. In many ways their encounter marks the culmination of three decades in which scientific studies of squids have exploded to the point that a pair of prominent marine biologists have called giant squids the 'marine versions of panda bears and dinosaurs' for the interest in biology they generate.[1] Previously, teuthologists worried that too many young marine biologists were choosing other directions, either because squids seemed less appealing than panda bears or because pelagic squids inhabited some of the least accessible areas of the planet. Kir Nesis, who revitalized squid research in the late twentieth century, proclaimed in 1982 that 'a continuous and rapid growth of interest in cephalopods – nautiluses, spirulas, squids, cuttlefishes and octopuses – has been observed recently.'[2] Fifteen years later, Malcolm Clarke, one of the heirs of Nesis's pioneering work, expressed a concern that too few marine researchers were specializing in cephalopod ecology: 'until two decades ago these numbered less than about 30 worldwide, while there must have been well over 50 times that number working on sea fish.'[3] Both of these scientists provide the same explanation for the new

One of the most striking discoveries of the German deep-sea expedition, *Vampyroteuthis infernalis*, the vampire squid from hell, as drawn by Carl Chun, *c.* 1900.

interest: 'decline in fish stocks, increase in the total value of cephalopod fisheries, and the realization that it is desirable to know more about fish, bird, and sea mammal food are now changing our attitudes towards cephalopod studies.'[4] The primary reason for the sudden interest in non-giant squids is not particularly comforting for what it indicates about the state of the world. As industrial fishing has depleted the populations of traditional species to virtual extinction, scientific attention has turned to squids, even to wonder whether they have always existed in such numbers or have filled the niches left by over-exploited species. The drop-off in populations of numerous marine species has several causes, but mostly the development of industrial fishing, which hauls in enormous catches, and, in raking up the seabed indiscriminately, turns traditional habitats into empty aqueous deserts. Modern marine biology has directed its attention cephalopodward because the previous lack of interest protected the squishies from enquiries into their potential value as exploitable resources.

Fishing for *chokka* (squid) off the coast of South Africa: squid fishing is international.

Clarke broaches that potential thus: 'we should ask which species of cephalopod might become of interest to man for food . . . there is clear evidence, mainly from estimated consumption of their predators, that there are large cephalopod resources in the open ocean available for exploitation'.[5] Just as natural historians relied on fishermen, modern biologists orient much of their investigations towards helping to make cephalopod fisheries more productive. Nesis makes this objective his primary justification for publishing a handbook on cephalopods when he states, 'high nutritional qualities, short life span, and an extremely rapid growth . . . make them highly promising animals for mariculture'. In this view, fishermen could give up their high risk adventures on the open sea to become tranquil squid farmers.

But biologists have found other squid values to exploit – even greater than food. Neurophysiologists have been attracted to the high teuthic intelligence, calling squids 'primates of the sea'. They stand out as potential resources for new and powerful neurological drugs to treat shock, cancer, and other human disorders.[6] In addition, scientists have found that the hard chitin of squids' beaks, hooks and sucker rings can be turned into durable prostheses for human amputees, and for biothermal plastic used in 3D printing.[7] Even if squid intelligence is not one most humans would sympathize with – as we do with that of gorillas, for instance, or cetaceans – it is recognizable in many ways. The physiological mechanisms supporting squids' predatory genius might not make great sushi, but they can interest a lucrative branch of the pharmaceutical and military industries. Scientists always need funding, and the promise of financial support from such well-endowed sources has sparked interest among young researchers.

Around the time of Steenstrup, two other marine biologists were also identifying deep-water cephalopods: the Englishman Willam Hoyle and the American Addison Verrill. Among them, these three scientists identified 115 cephalopod species during the seven-year period from 1879 to 1886. In 1898 the German marine biologist Carl Chun led the German deep-sea expedition aboard the ship *Valdivia*, and explored the deep waters around Antarctica. In 1910 Chun published his beautifully illustrated work *The Valdivia Expedition*, which included the first reference to (and illustration of) *Vampyroteuthis infernalis*, in the naming of which he showed more wit than his Danish predecessor. Cephalopod research surged for two decades after Chun's work, with 83 new genera and 303 new species of cephalopods identified. And then research fell away until Nesis was able to announce renewed interest in the 1970s.[8]

The latest version of the primary taxonomic catalogue of cephalopods, *Cephalopods of the World*, has expanded to three volumes, allowing one to be devoted entirely to squids. The newer generation of editors, Patrizia Jereb and Clyde Roper, observe that 'during the 20-plus years separating the two editions, the rapid development of cephalopod fisheries worldwide and the simultaneous increase in the population of fisheries scientists, their research and publications, made available an enormous amount of new data and research results.'[9] The increased attention to cephalopods generally can be seen in the trebling of volumes in this catalogue; and the interest in squids particularly is made clear simply in the fact that the volume devoted to them is almost twice as thick as the original single volume that included all four cephalopod orders (as they were then defined).

Correspondingly, modern teuthology is anything but a singular enterprise. The plethora of squid genera and species has compounded the general uncertainty among older natural historians. A simple definition of 'squid' has become virtually impossible, so that biologists encounter vastly different environments, morphologies and behaviours. While some scientists direct their study towards potential for exploitation, other scientists focus on teuthic adaptability, communication and intelligence. The exploitative investigations generate a lot of wealth and objectify our fellow creatures, and at the same time produce an overwhelming amount of information. Both the investigations that aim to turn squids into gold mines and those that look for ways to engage with these fascinating and strange beings reflect human attitudes towards animals generally. Both warrant consideration as reflections of cultural attitudes towards squids.

Aristotle divided squids into the common and easily seen inshore varieties, and the offshore varieties that were seen less often and could only be guessed at. Modern teuthology, with its global reach, faces a similar limitation. Biologists admit that most of their knowledge about cephalopods is based on the species inhabiting the continental shelves close to shore, along with the offshore species that migrate into shallower waters. This means that scientists have to base their understanding of cephalopods generally on roughly 15 per cent of all species. Squids inhabiting areas beyond the continental shelves often move through waters too deep or stormy to be easily fished or studied. Because modern research on cephalopods has increased so dramatically, classification systems remain in flux. 'The classification above the family level is controversial,' say Jereb and Roper, but such a situation 'is not unexpected for a group of organisms that has undergone explosive research attention in recent decades'.[10]

The terminological slipperiness that led the previous two millennia to refer to cuttlefishes, polyps and calamaries interchangeably has given way to a hyper-precision in cephalopod taxonomies. The systematist G. L. Voss complained in 1977 that earlier generations of taxonomists had 'abhorred specimens without names and created new genera and species wholesale'. Voss raised this concern due to the fact that at the time 85 per cent of cephalopod genera contained fewer than five species, and nearly 50 per cent contained only one. This overwhelming proliferation of categories threatened either to create taxonomic mazes of virtually indeterminate relations between species and genera, or to destroy the conception of species identification. The resolution Voss came up with was to recognize that 'with few exceptions, monotypic genera consist of oceanic midwater species while polytypic genera are represented by shallow-water continental shelf or deep-benthic forms, both areas apparently conducive to a high degree of speciation.'[11] Among squids, this broad spatial division was subsequently refined into the two groups of myopsid and oegopsid squids, corresponding generally to Aristotle's division of *teuthis* and *teuthos*.

A number of squid groups – such as bobtail squids, bottletail squids, pygmy squids and ram's horn squids – are all categorized nowadays as cuttlefishes. (Another worthwhile distinction is that 'squid' and 'cuttlefish' officially refer to individuals or to a

Bobtail squids reflect the uncertainty in classifications of cephalopods. Currently they are considered to be closer to cuttlefishes than to squids.

single species, while 'squids' and 'cuttlefishes' refer to two or more individuals or species.) The most illustriously lugubrious former squid, Chun's vampire squid from hell (*V. infernalis*), now owns its own order, Vampyromorpha, though teuthophiles the world around will forever cling to its squidly identity. This frightening-looking creature inhabits some of the darkest depths of the ocean, and is thought to mark the evolutionary point where octo-pods and decapods diverged. This cephalopod, about 25 cm (10 in.) long – too small to count as a kraken – has eight legs and, instead of the two tentacles distinguishing squids and cuttlefishes, has two long filaments that can be retracted. Its reddish-black skin, blue eyes and webbing between the appendages recall the caped characters of Saturday matinees, and the depth of their habitation makes them truly infernal.

This oral view of *Berryteuthis magister* shows the beak, which modern scientists rely on for species identification.

One of the most significant developments in cephalopod systematics came in 1986 when Clarke published his *Handbook for the Identification of Cephalopod Beaks*, which made an extensive classification of squids possible, along lines similar to the Linnaean classification of flowers according to the sex organs. The problem long confronting systematists of cephalopods was to find one common physiological feature – like flowers' sex organs – that differed from species to species. Just as Steenstrup proved that the monk fish was a squid by displaying the monk's beak, modern taxonomists have used beaks since 1955 to identify secular cephalopods eaten by sperm whales, pinnipeds and birds. Since the rest of a squid's body consists almost entirely of soft tissue that breaks down quickly and entirely, the realization that the chitinous beaks resisted digestion, and were distinct among species, made classification meaningful for the first time. Not surprisingly, the enormous number of cephalopod orders, families, genera and species posed the daunting challenge of organizing a taxonomy whose usefulness depended on general acceptance of nomenclature. As indicated by Aristotle's tantalizingly vague distinction between two squid groups, a confusion of 31 families, 95 genera and 450 species (the numbers appear to keep growing) seemed almost certain. So, as Clarke relates, 'the time was right' to develop a common methodology for describing beaks, to revise existing taxonomies and to set directions for future research based on the new taxonomy.[12] A systematist would often be able to identify the genus or species of a beak based on any soft tissue still attached to the beak (and such a detail would also indicate the likely location where the specimen had been eaten, since the incomplete digestion of the tissue would mean the individual had only recently been swallowed).

Clarke lists four principal parts of the lower beak on which he bases his categories: the *rostrum* – or the point that would tear

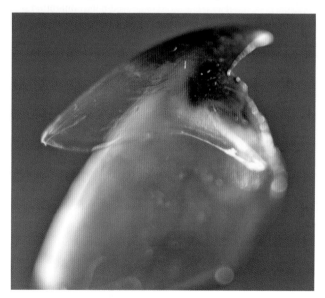

The beak of one of the better known squid species, *Loligo vulgaris*, the common squid.

the flesh of the squid's victim; the *hood* – or the thick structures supporting the rostrum beneath and along the sides; the *wings*, which serve as the working jaws; and the *lateral walls*, which run behind the wings and downward from the hood to give the entire structure its strength.

A look at even this half of the overall beak structure makes it clear that this is an organ for cutting and tearing, serving to rip the victim to shreds quickly. And within the beak the radula, a toothy tongue that can hold as many as seven rows of long, pointed teeth angled inwardly, continues breaking the food down into digestible bits; the dentine angle ensures that the captured prey will move in only one direction: inwards. Altogether, the beak, the radula, the muscles that move them and the salivary glands constitute the buccal mass lying at the centre of the arms. The buccal membrane – the muscular system that moves the beak – may contain its own set of suckers, lined with chitinous rings that often have either serrated edges or teeth to ensure that the prey may be held steady while the beak tears its flesh and the radula pulls the torn bits inward and tears them down even further. The beaks of some species are said to carry venom, though most of the problems

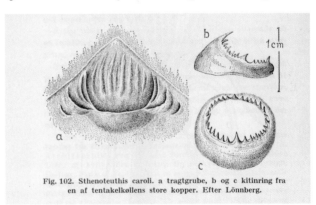

Fig. 102. Sthenoteuthis caroli. a tragtgrube, b og c kitinring fra en af tentakelkøllens store kopper. Efter Lönnberg.

Squid sucker ring 'teeth'.

experienced by humans bitten by squids have been caused by secondary infection. Nonetheless, any squid possesses at the centre of its being a formidable system for eating a victim's flesh as quickly as possible.

The mouths of squids are large compared with the mouths of fishes.[13] In contrast, the oesophagus has a small diameter, since it passes through the brain. This means that, even though squids (and cephalopods generally) seize rather large prey, they have to hold on to it and tear off small bits of soft tissue at a time. As one of the squishy creatures, a squid would logically have no appetite for hard, bony food. This fact creates problems for the scientists who seek to know exactly what squids eat by cataloguing the contents of their stomach. Squids prey on almost anything that does not eat them first, and they hunt continually, even after being netted. The stomach contents of even freshly caught squids do not necessarily reflect what they would have eaten outside the net, especially given the indiscriminate sweep of enormous modern dragnets.

The head bears the two tentacles and eight legs, and provides the modern Greek word for the entire class, Cephalopoda, or 'head of feet' (the self-identification of cephalopod fans as 'ceph-heads' does not quite catch the point).[14] Within their leg-ringed heads, squids possess sizeable brains, an important detail for understanding their abilities to move swiftly, to see in the gloomy

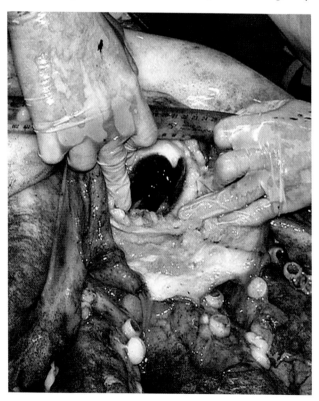

Measurement of the size and shape of the beak is used to distinguish between the many species of squid.

aqueous environment and to change colours – all qualities of a highly effective predator. Unlike those of octopuses and cuttlefishes, squid heads appear only slightly connected to their mantles (as though they were wearing their mantle like a mitre), but in fact they are linked by the 'nuchal-locking apparatus'. This cartilaginous structure enables a squid to manoeuvre its mantle and head in more fluid and rapid motions than other cephalopods, making them faster and more effective predators. The cartilage is surrounded by a series of folds running posteriorly from the back of the head – the nuchal crest – underneath the foremost part of the mantle. This area contains another of the many sensory organs possessed by squids, and apparently by all coleoid cephalopods. It appears to perceive light, but of what sort, and how it would differ from the eyes, no one knows.

The hydrodynamic mantle of squids generally appears tubular (or sleeve-like), also distinguishing them from cuttlefishes and octopuses, and is given its shape by the cartilaginous structure running dorsally through it, known as the gladius or pen. The mantle lies behind the head, and holds the major organs, such as the gills, three hearts, the anus, as well as the long cavity through which the squid compresses water that it shoots out through the funnel for jet propulsion.

Recent taxonomic divisions have affirmed Aristotle's basic division and organized squids into Myopsida, whose eyes have a cornea-like covering, and which mostly consist of inshore (neritic) families, and Oegopsida, whose eyes lack that covering, and who live in the deeper waters. Myopsid squids lack hooks on their tentacles (though they may have sharp teeth on their clubs), while many of the Oegopsida possess them. So in the sub-class of Coleoidea – defined by having either an internal or absent shell, chromatophores, ink sac, large brain and eyes, thus including octopuses, cuttlefishes and squids – there is the order Teuthoidea,

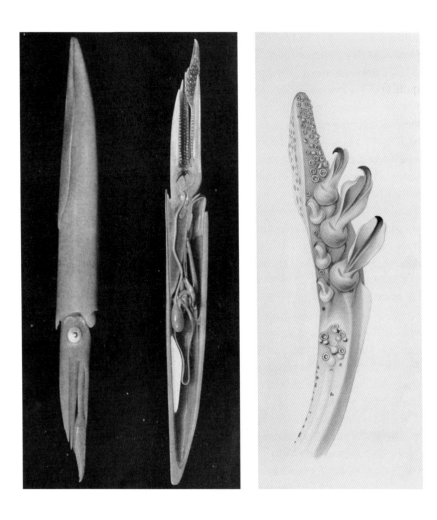

which contains the two sub-orders of squids, Myopsida and Oegopsida. The inshore Myopsida consists of two families, Loliginidae and Australiteuthidae, species from both of which are quite familiar to natural historians and to diners around the world. And the deeper-going oegopsids consist of twenty or more families, many of which have not been studied adequately enough to dispel the vagueness of Aristotle's account of *teuthos*.

This vast and dizzyingly diverse group of animals evolved from early molluscs by getting rid of their shells. Cephalopods first appeared in the Cambrian oceans, some 500 or 600 million years ago, making them genuinely ancient creatures. The 'modern' coleoids are generally believed to have appeared sometime during the Jurassic period, which lies in the middle of the Mesozoic era, or about 150 million years ago. Some palaeontologists date coleoids much earlier, however, going all the way back to the Devonian era of 350 million years ago.[15] Egbert Giles Leigh argues that these ur-cephalopods produced 'both straight-cone and coiled' shells, providing the first real division of consequence. For some of the straight-cone cephalopods crawled while others swam swiftly; the coiled cephalopods also swam, but more slowly. The crawlers gave way to gastropods (early snails, or slugs). The

This model, in the American Natural History Museum, clearly illustrates the location of the various organs inside the mantel of a common Atlantic squid.

Chun's drawing depicts the hooks that distinguish tentacles from arms of *Abraliopsis morisi*.

Reconstruction of a Precambrian-era proto-squid in the Museum of Natural History, University of Michigan.

coil-shelled cephalopods endured all the way through the Eocene period of 50 million years ago, and then went extinct. It was the straight-cone swimmers who gave way to squids, according to Leigh.[16]

The Mesozoic group known as belemnites possessed an internal shell that could be either straight or conical. In some of these ancient creatures, the internal shell chambers moved closer and closer together, and eventually flattened to form something like the cuttlebone of the sepias. Other belemnites let their shells lighten even more to remain as only the quill-like cartilage of squids.[17] Some palaeontologists theorize that competition with fish and marine reptiles in the shallow waters pushed cephalopods into the deeper areas off the continental shelves. In such depths, the hydrostatic pressure caused difficulties for interior shells that contained low-pressure gas chambers to facilitate flotation. The evolutionary resolution of this problem lay in shedding the shell in order to swim in the depths. And without their shell, these creatures could swim well enough to return to the shallows to take their revenge by competing with fish as the far more effective predator. This theory would explain the division between the myopsid, inshore squids, and the oegopsid squids that reside in the deeper waters offshore.

Clearly, coleoid development really began in a recognizable fashion when the ancient creatures rid themselves of their shells. With this point, teuthologists open an interesting area of speculation. Cephalopods, they remind us, are molluscs, and yet they behave like teleosts, which is to say bony fishes. From early in their long evolution, cephalopods have demonstrated an adaptability, being able to change themselves in accord with circumstances or opportunities. After the Devonian extinction, when a large percentage of earthly life disappeared, animals repopulated all areas of the world prolifically. An exception was in the deepest oceans, where, speculation has it, low oxygen levels precluded animal habitation, except for cephalopods, who seem to have adapted just fine.[18] This potentially limitless development has everything to do with getting rid of the molluscan shell in order to perfect their predatory squishiness, and could in the looming future

Small belemnite fossils, mostly from Europe.

enable them to endure climatic changes threatening most other life forms.

While the myopsid squids would include species most commonly encountered by humans, the deep-water oegopsids include a much more diverse array of species spread throughout the oceanic world. Myopsids all possess the regular span of legs along with a pair of clubbed tentacles that contain suckers. Of the two myopsid families, the first, Australiteuthidae, contains the single species of *Australiteuthis aldrichi*, the Australasian inshore squid. This is a small squid, around 2 cm (1 in.) long, living in the waters along northern Australia and Papua New Guinea. This species is known from only about three dozen specimens, though as of 2009 no living members of Australiteuthidae had ever been observed.[19]

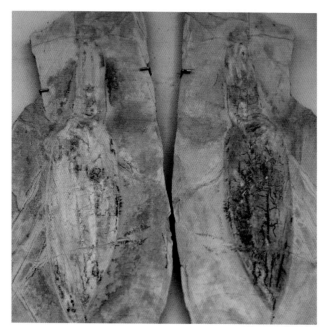

Fossil of *Donovaniteuthis*, an extinct squid.

Such a tantalizing situation of knowing a species exists without being able to find it alive is common in teuthology. And even the remains of squids provide little information, for, being as soft as they are, they will lose many of their distinctive characteristics quickly upon dying, being washed up on the shore or being caught in a net, thus remaining aloof from scientific interrogation.

A mating pair of *Euprymna tasmanica* (dumpling squid), recently considered to be closer to cuttlefishes than squids.

The second family, Loliginidae, contains ten genera, nine sub-genera and 47 species found in near-shore waters worldwide, in temperate, tropical and subpolar regions. The delineations of the genera still remain uncertain, which is not surprising for such a widespread family of creatures. But the Cephalopod International Advisory Council held a workshop in 2003 in Phuket (Thailand) to attempt a resolution of some classifications. What we can say is that all these squids live close to shore and along the sea floor. But that general description covers a dizzying variety. Some species even tolerate low salinity, such as *Loliguncula brevis*, found in the

Loligo squid off the coast of Japan.

western North Atlantic. This is one of the many details suggesting that opportunistic squids are exploiting the changing seas to take advantage of decimated populations of other creatures – such as their predators – and adapting to newly possible habitats.

The southern Australian fishery is the most productive in the world, and consequently the one most studied by marine biologists. Scientific attention has shown that catches of southern reef squids, *Sepioteuthis australis*, contain notably higher portions of males than females, indicating that fishing during the spawning season has disrupted the sexual selection, and could consequently suppress the genetic diversity of these squids.[20] This is an important recognition that, like the cod and so many other creatures once seen as limitless resources, squids can be depleted. The symbiosis between teuthologists and the fishing industry has led to regulating the season of the squid fishery to avoid the spawning time.

One of the most common species is the bigfin reef squid, *Sepioteuthis lessoniana*, whose range extends throughout the Indian Ocean, into the Pacific, up to Japan and over to Hawaii.

This squid has the appearance of an oblong disc, as its fins take up almost the entirety of the mantle length, and are up to three-quarters as wide as they are long. The bigfin reef squid can grow to a mantle length of about 45 cm (18 in.), with long tentacles each bearing a club with toothy sucker rings. When the water begins to warm in the spring, bigfins move shoreward to form close pairs for mating, and the males flash antagonistic colour patterns at their competitors (and alluring patterns at their sweethearts). Biologists express particular interest in these squids, not so much because of their social life, but because they tend to grow and mature quickly, making them attractive as laboratory specimens and as possible stock for squid farms. The biologists point out that the squids can be induced to eat processed pellets, and that so far the domestic bigfins have restrained their cannibalistic tendencies from the wild. And most happily of all, they have willingly engaged in reproductive activities.[21] Industrial squid farming lies clearly in reach, making some species an even more common part of humans' diet than they already are.

Sepioteuthis sepioidea off the coast of Bermuda.

This small selection from the fifty named species of myopsid squids shows the variety and wide distribution of inshore squids. These species have long dominated the interest and knowledge of humans because of their habitation close to shore, making them easily observed, caught and eaten by coastal communities. Because the oegopsid squids live beyond the continental shelves, they remain far more elusive, despite their greater size and numbers. These squids appear in almost all seas and oceans (the only exception being the Black Sea, impoverished of squids), and occupy virtually all depths, from the surface and midwater, down to bathyal depths of more than 4,000 m (13,500 ft).

Foremost among the oegopsids, the giant squid (*Architeuthis dux*), named by Steenstrup and filmed by Kubodera, is indeed large: reportedly up to 18 m (60 ft), including its tentacles, and with its mantle measuring 3 m (almost 10 ft). But it is always worth remembering that giant squids are rarely seen by humans, and individuals of such size are encountered even more rarely. They are said to weigh in at around 500 kg (1,100 lb) outside the water. Even after Kubodera's video showed a sinuous and slow-moving squid, a debate has continued over whether these are sluggish creatures or ones capable of monstrously aggressive attacks.

Giant squids are among the ten families that achieve a neutral buoyancy – the ability to resist sinking or rising uncontrollably – through ammonia chloride stored throughout the mantle. The ammonia, which is less dense than seawater with its heavy sodium ions, neutralizes squid weight and the density of squid muscle mass. This element makes these squids unsavoury to many predators, though sperm whales have clearly developed a taste for ammonia, as they famously eat considerable quantities of *Architeuthis*, leading fabulists to fantasize about the battles between deep-sea monsters. Most squids in the less spectacular family of Cranchidae, commonly called glass squids, also rely on the ammonia sac for their neutral buoyancy – in fact, the mantle cavity is divided into two areas, each containing ammonium chloride. Different species distribute the ammonia in different parts. The large cranch squid genus (*Megalocranchia*), for example, keeps it in a sac towards the rear of its transparent mantle, while the graceful long-armed squid (*Chiroteuthis veranyi*) holds the ammonia in its head, and as a result hangs mostly head downward, with its sinuous arms and tentacles flowing towards the sea bottom. Other squids, including *Architeuthis*, distribute the ammonia throughout their muscle-tissue. In contrast the Bathyteuthidae – mostly small, deep-diving squids – contain an oily substance in their liver and around their mantle and head that includes some as-yet-unidentified cation that neutralizes the heavier anions of the seawater.[22] A recent study has found that even the fast-swimming, negatively buoyant species, such as *Loligo vulgaris*, still possess very high quantities of ammonia, held to be a key reason for the rapid decay of squid bodies and the paucity of teuthic fossils compared to those of other cephalopods.[23]

The colossal squid, *Mesonychoteuthis hamiltonii*, comes pretty close to equalling *Architeuthis* in size, at around 3 m (10 ft) and 450 kg (1,000 lb). The colossals have not commanded the fame of the giants, possibly because they swim through the Antarctic waters, staying beyond latitudes of 70 degrees south. Like the giants, they are preyed upon mostly by sperm whales; unlike *Architeuthis*, they do not maintain buoyancy with ammonia chloride, and so have attracted the attention of the United Nations biologists as a possible food for humans, if only the whales can be restrained, and the stormy waters of the Southern Seas calmed.

Squids interact with currents and changes in light, heat and salinity by taking on varieties of migrations. Almost all species migrate vertically throughout the daily cycle, staying in the deeper waters during the daylight hours, then moving upward at night. This diel migration enables squids to hide from overhead predators (birds) during the day, and to conduct their own predations in the safety of darkness concealed from other predators below. The depth to which a squid descends varies according to species and location. Since myopsid squids occupy the continental shelves, their vertical migrations are limited by the shallow waters, while the oegopsids can often migrate close to 1,600 m (1 mi.) in depth. Nonetheless, some myopsids, such as the slender inshore squid (*Doryteuthis plei*) that inhabits the Atlantic coast along the Americas, stay along the bottom during the day and then move right up to the surface at night. Horizontal migrations – in which entire populations of a species move from one region to another – occur among myopsids seasonally, often due to spawning. The European squid (*Loligo vulgaris*) living in the eastern Mediterranean migrates to deeper offshore waters in late autumn, then the larger individuals will swim inshore in January to spawn, followed by the smaller individuals in the summer. This pattern raises the possibility that the two types of squids identified by Aristotle were actually the same species at different stages in their life cycle.

As with virtually all aspects of squid life, however, knowledge of the basic facts about migration has its limits. Different populations of the Patagonian squid (*Doryteuthis* [*Amerigo*] *gahi*) inhabit the Pacific waters along the western coast of South America, and around Cape Horn northward, past Argentina and the Falkland Islands (the Malvinas). Only the Falkland population has been studied, due to its importance for fisheries in the area. Of these squids it can be said that they spawn in shallow water, then migrate down the continental shelf after hatching. During their maturity they commonly live at depths of close to 300 m (1,000 ft), but can dive down to double that. Within this migratory pattern, differences occur between sexes and between larger and smaller individuals. Squids of about 10 cm (4 in.) mantle length will stay at depths of about 120 m (400 ft). Males with slightly longer mantles, say 11 cm (4½ in.), will range from about 170 m to 250 m (560–820 ft), while females of the same size will stay deeper, from 250 m to almost 300 m (820–1,000 ft). The larger, deeper-swimming males move to waters as shallow as 30 m (100 ft) to wait for the females, who may linger until the middle of the spawning season to join their eager counterparts.

As the oegopsid squids live in the deeper waters, they naturally conduct longer vertical migrations, though not always on a daily basis. Many glass squids move gradually deeper as they pass through progressive stages of maturation, ultimately taking up residence in depths sometimes around 2,400 m (8,000 ft). Glass squids inhabit such large areas of the world's oceans that horizontal migrations are not so apparent. The large cranch squid (*Megalocranchia*) in the South Atlantic from the Cape of Good Hope to Argentina is thought to migrate upwards at night to depths between 30 m and 700 m (100–2,300 ft), preferring to linger in the 2,400 m (8,000 ft) range to avoid light during the day. These large squids can commonly reach a mantle length of 175 cm (70 in.).

The rather stunning jewel squids – distinguished by the large number of photophores throughout their mantles, tentacles, arms, heads and dazzlingly around their eyes – show up throughout the world, and migrate vertically, though not over great distances. The flowervase jewel squid, found in the tropical and sub-tropical waters of the Pacific and Indian oceans, congregates in a range of 450–700 m (1,500–2,300 ft – the upper bathypelagic zone) during the day, then swims up to around 150 m (500 ft) depth at night. One of the intriguing features of the flowervase jewel is its dimorphic eyes: the left eye tends to be large and point upward, while the smaller right eye points down and back; this squid can thus make use of the dim light that penetrates to its depth during the day, while also perceiving bioluminescence of prey or predators below.

The abundant family of Ommastrephidae are probably the most migratory of oegopsid squids, as they move about vertically and horizontally. Because of their large populations and their impressive migrations, these squids have been found to play a central role in the oceans' ecosystem. As they consume their prey in one area, then move to another area to be eaten by other predators, ommastrephids effectively transfer large amounts of biomass from one area to another. The northern shortfin squid (*Illex illecebrosus*) mostly lives in the western Atlantic along the continental shelf and upper slope from Canada to Florida. When this squid spawns off Cape Hatteras, the rapidly moving currents of the Gulf Stream transport the paralarvae northward to Nova Scotia and Greenland. Teuthologists speculate that the interactions between the warm, northward currents of the Gulf Stream and the slope-water thermostad of about 150 m (500 ft) in depth, where the temperatures remain around 13–18°C (55–60°F), provide the fertile hunting grounds for this squid.

So vital to the overall life of the ocean are the shortfins that biologists describe them as a 'biological pump' transferring energy and protein from their feeding grounds to their spawning grounds, over 1,000 km (620 mi.) away.[24] Because they grow more quickly than do their potential predators, the northern shortfin squids survive the stages of life in large numbers. They not only consume a large number of other creatures, but, in the way of squids, convert their food into body mass quickly. In the warm waters of the Gulf Stream, food tends to be scarcer than inshore, leading the squids to move to the cooler slope waters where they do most of their growing. When they attain their mature size of about 35 cm (14 in.), they migrate south to spawn. By this time, they have also amassed a goodly amount of lipids in their digestive gland which help their expenditure of energy during the long migration in late autumn. Thus, the image of the 'biological pump' describes the movement of paralarvae northward, the consumption of large amounts of protein in the northern waters, and then the migration of adult squid southward to spawn, die and decay. The regular sequence of the consumption and movement creates the impression of a systematic operation in which squids serve the singular function of moving protein up the food chain.

One part of this chain is the human population that relies increasingly on reliably large catches of protein-rich squids. A key element of the relationship between marine biologists and fishermen lies in developing the accurate knowledge of squid behaviour in order to ensure a sustainable catch. In a reversal of the arrangement between fishermen and natural historians, the modern fishing industry turns to the scientists for information about squid behaviour that might facilitate the largest possible hauls. At the same time, marine biologists seek to ensure a sustainable catch of wild squids or to develop reliable domestic species.

At least one species gives indications of exceeding human management, however. One of the best-known squids of our modern age – rivalling even the giant squid – is the Humboldt squid, or jumbo flying squid (*Dosidicus gigas*), which is generally 0.9–1.2 m (3–4 ft) long. As its name indicates, this squid inhabits the Humboldt and California currents that move along the western coast of the Americas. Its fame, or infamy, comes from its highly aggressive hunting behaviour, and its tendency to rove in predatory groups. The Humboldt has such a strong attack behaviour that it has been described as attempting to prey in significant numbers on other Humboldts that have been jigged, so that a 'squid ball' forms of layered squids eating those beneath. Not only do they become prey for each other, but they are also preyed upon

The beak of the ferocious Humboldt squid (*Dosidicus gigas*).

in the 'biological pump' by the fearless sperm whales that cruise the Humboldt current, along with beaked whales, bottlenose dolphins, blue marlin, swordfish, sailfish, mahi mahi, yellowfin tuna and hammerhead sharks.

With such a long list of predators, Humboldt squids are perhaps fortunate in commanding one of the largest squid populations in the epipelagic zone – the uppermost layer of water extending down to about 200 m (650 ft). The elimination of other predators in the area, those that would both compete with and eat the Humboldt, has led to a population explosion. Consequently, the Humboldt has been identified as one of the two causes – the other being over-fishing by humans – for the precipitous decline in the population of Pacific Hake. The industrial fishery for the Humboldt itself grew very rapidly for three years, from 1978 to 1980, and then collapsed. When the fishery moved on to other species, the Humboldt population recovered sufficiently to attract the fishing industry again, and to such an extent that by the second decade of this century the massive fishing fleets could be seen from space. The dramatic rise in Humboldt populations also led to several mass beaching events, resulting in large numbers of rotting squids offending the vacationers.[25]

Because of the fairly nascent aspect of squid research, one question raised by dramatic population shifts is whether these are a relatively common characteristic of squids or one more sign of a notable change in the oceanic ecosystem. Recently, other squid populations have demonstrated similar dramatic and sudden increases, and indeed scientists have wondered if we are witnessing a cephalopod explosion throughout the world. *The News* from Karachi (Pakistan) reported on 28 September 2017 that large aggregations of purpleback flying squid (*Sthenoteuthis oualaniensis*) had appeared in the Arabian Sea from November 2016 to January 2017. The article attributes the sudden explosion to climate change, and compares it to the jellyfish blooms that have increasingly occurred worldwide.[26] At almost the same time, in the UK, *The Sun* reported a 'bonanza' harvest of cuttlefishes off the coast of Devon. Most of this harvest, the paper explained, was sent off to China, since the English palate does not knowingly savour this cephalopod.[27] The explosions in both cases attracted such sizeable fishing fleets from around the world that, as with fleets pumping up the biomass of Humboldts, they have been photographed from space.

The reason these fishing fleets become visible to space travellers is that they rely on lights to attract the squids upward and towards the hooked jigs. In the mechanized fishing of today, the lures hang from lines extended from large winches; the drums of the winches have an oblong shape to make the jigs halt and jerk in the water to simulate a living creature squids might want to eat. These fishing strategies exploit the visual acuity of squids, as well as the peculiar orientation of their eyes that would serve predatory habits. Along with their swiftness, and their fluid motility, squids seem to arouse human interest because of their large eyes – eyes that, unlike those of cuttlefishes or octopuses, resemble our own eyes with their round pupils.

But squids live in water, where light moves differently than in the atmospheres humans inhabit, and in their deeper ranges little light penetrates at all. What a squid actually sees with its large eyes therefore becomes an intriguing mystery of its own. Prominent cephologists Peter Boyle and Paul Rodhouse have used their own scientific vision to examine the differences between cephalopod and human eyes in order to lay some groundwork for speculation about squid vision.[28] Like us, cephalopods

The lights of squid boats fishing off Hakodate, Hokkaido, Japan.

perceive images, a quality lacking in other molluscan vision. Among vertebrates the back of the retina is lined with receptor cells shaped as rods and cones that each serve to receive information about movement or form, and that work differently in low or high light. In cephalopod eyes receptor cells also line the retina, but these cells have a wholly different structure: from each cell rhabdomeres, or rods, project towards the incoming light. These rhabdomeres are aligned equally, half vertically and half horizontally, while human rods and cones have a much more random arrangement. Further, in vertebrates the receptor cells have a layer of other nerve cells lying over them, which are absent in cephalopods, meaning that the visual image falls straight onto the receptor cells, which are in turn connected directly to the optic nerves passing into the brain.[29] As a result, squids receive visual images more directly than do vertebrates.

In addition, the alignment of the rhabdomeres makes squids particularly sensitive to vertical and horizontal configurations. More specifically, squids have an acuity for polarized light, so that they perceive light configured to individual planes – vertical and horizontal – and thus can distinguish among different polarities, which humans cannot do. Polarized light is not generally perceptible to humans, nor to vertebrates generally. A beam of the non-polarized light, which we can see, consists of rays vibrating in all orientations, while polarized light vibrates in a single plane. Polarized vision contributes to the many other abilities to make squids effective predators. For example, the scales of fish reflect the same amount of indirect light as the light shining directly behind them, providing them with a camouflage effective against most predators. But their scales also produce some polarized light, which makes them visible to squids. While squids are generally believed not to perceive colours, some researchers have suggested that polarized vision is analogous to colour vision.[30]

Apart from their impressive eyes, squids possess photosensitive organs both inside and outside their bodies that do not perceive images so much as gradations of light. Although teuthologists have noted their presence, they remain mostly baffled by them. Squids see in multiple ways that befit a swift-moving and voracious predator passing through the different layers of the water column, where light varies considerably. Squids engage with a different range of light, and in different ways than terrestrial humans do. The squid world is not the human world, and to contemplate a squid's engagement with the watery environment challenges our assumptions of what we take as accurate and comprehensive perception of a singular world.

Along with their aquatically specialized vision, squids possess a pair of statocysts – complex structures that facilitate their bodily orientation as they move through the water. These structures consist of a statolith – which, as the name suggests, is a stony or calcareous particle – suspended on ultra-thin cells (the macula) in a fluid-filled cavity. As gravity pulls the statolith one direction or another, the movement is detected by the macula, which orients the squid directionally. Orientation can work up and down, side to side or in any other direction. The statocysts have been called 'the most complex of all invertebrate receptor systems', rivalling the vestibular system in vertebrates for sophistication.[31] Scientists have recently learned that statoliths grow in daily increments, and can thus be used to determine an individual squid's age.[32]

The statocysts functionally resemble certain structures of the human's inner ear that facilitate balance and hearing. But while squids have no organs obviously analogous to ears, they do possess a series of skin cells along their head and arms oriented and polarized in ways that make them capable of detecting vibrations in the water. This sense, known as the lateral-line analogue, enables a squid to perceive movement as small as

six-tenths of a micrometre. At such an acute sensitivity, squids are said to be capable of detecting movement over 25 m (85 ft) away. The lateral-line analogue has focused a scientific debate over squids' perception of sound. Some take the analogue as the equivalent of hearing, while other scientists suggest that the 'deafness' of squids enables them to 'cope with some of their most dangerous predators, the highly vocal cetaceans', and yet other scientists suggest that it is specifically through intensely focused sound that cetaceans stun squids in order to attack them. Another group point out that sounds have been used effectively by commercial fisheries 'to enhance squid capture by day or under a full moon'; a useful strategy when the customary lights lose their efficacy.[33]

Long-time teuthologist Jennifer Mather has focused on squid intelligence and their communications, setting out a rudimentary

School of *Sepioteuthis* in uniform dark coloration.

grammar of squid movements and postures.[34] She recognizes that without a fixed skeletal system, squids are capable of a corporeal flexibility to move in an overwhelming variety of expressive 'actions and postures'. She then works to catalogue and categorize the movements into 'stereotypies that reduce the variation' into recognizable and readable codes.[35] Mather and her colleagues observed Caribbean reef squids in a single location, off the island of Bonaire, for five years, and outlined the 'stereotypies' of squids' arm and tentacle positions, assigning them names, such as 'standard', 'extended', 'belled', 'devilish' and 'bad hair'.[36] Mather does not venture to translate what a 'bad hair' posture would mean in a human, but it seems fair to assume that conversations among intelligent beings who live in a world uninhabitable to us, and who perceive that world in ways unimaginable by us, would be difficult to translate into most human idioms.

In another study of squid talk ('squiddish'), Hanlon and Messenger make the utterly captivating comment that 'signals may be either honest or dishonest.'[37] Their report offers no speculation about what 'dishonest' messages might entail, or what they might suggest about the squid choosing to lie. Nor does it venture any thoughts about ways a squid interlocutor might respond to a statement revealed to be a lie, or half-truth, or even a slight exaggeration. The possibility of dishonesty – or twisting the truth – suggests that squid communication works at a level more sophisticated, or perhaps more playful, than a simple exchange of information. Could cannibalistic predators employ 'devilish' or 'bad hair' postures as jokes? Could it be that squids play, have a squishy sense of humour?

These questions are not as far-fetched as they might seem. In a 2014 article, the team of Michael Kuba, Tamar Gutnick and Gordon Burghardt present their study of playful cephalopods. They confine their observations to octopuses, since squids choose the deep ocean for their playground, beyond the reach of human interpreters. Yet the aim is to establish octopus behaviour as a basis that can be extended analogously to squids and cuttlefishes. Combining the behaviour of exploration with that of play, these biologists begin by quoting the opening words from a study published almost forty years earlier: 'Who would dare study play?' Such a teasing question suggests that marine biologists themselves have a devilish sense of humour (and notoriously bad hair). And even though this study proceeds by formalizing play into three types – locomotor-rotational play, object play and social play – the scientists' wry wit reveals itself when they observe that the octopus, with its tendency for cannibalism, 'cannot be expected to engage in social play'.[38]

These studies from the past decade might indicate something of a turn. If the research into squids is undergoing a real expansion – because of the depletion of other oceanic life forms, the discovery of cephalopod comedy, or any other reason – the teuthologists are clearly asking new sorts of questions that go well beyond classification of species and population counts. In the 1970s systematists like Gilbert Voss said that squid science was still in its alpha or descriptive phase. These recent studies suggest a move into a different level of inquiry. Questions about squid language, intelligence and predatory jokes open unexpected possibilities for appreciating the strangeness of the creatures with whom we share the world. In 2013 Mather made the strong comment that 'just because an animal does not have similar brain circuitry to ours (homologous) does not mean that it cannot have that function (analogous).'[39] Squids, she reminds us, exist 'worlds away from us', so, even if we cannot recognize any aspect of their predatory intelligence in ours, even if we do not find their version of comedy amusing, we can employ our own sophistication to approach squids as creatures possessing highly developed capacities suited to their world, not ours. That is what makes them endlessly fascinating.

Marine biology has developed a wealth of knowledge about squid distribution in the world's seas and oceans. It has detailed the differences among the genera and species, and identified various ways squids perceive their world and communicate with one another. Since the time of Nesis, only thirty years ago, the study of squids has expanded significantly, and in broadly different directions. And, like Aristotle, Pontoppidan and Steenstrup before them, modern teuthologists gather information wherever they can, and factual knowledge will always be important. As Hanlon and Messenger say, 'we know so little about cephalopods that almost any evidence is worth listening to.'[40] They make this point by way of stressing the importance of listening to fishermen's account of the creatures they come into contact with. Fishermen's tales, along with legends and myths – the subject of the next chapter – offer a completely different accounting of squids than factual scientific studies do. In the cultural record they provide of human attitudes towards squids, they can enable us to develop a more speculative branch of thinking about squid intelligence and wit. Such thinking could possibly open the way towards contacting an alien life-form in our own world.

3 Folk Tales and Legends

Legends of squids have sustained a tradition parallel to and inter-twined with efforts to analyse and classify squids. Following squid lore takes us quickly from the taxonomic discipline of concrete certainty to hazier, squishier, more fluid representations of squids as unseen, and generally nebulous, sea monsters. It is safe to assume that the legends began with fishermen or other travellers over the sea, but they have always spread to land, where they have inspired visual and literary depictions. Modern teuthologists have tried to guide fishermen on industrial ships to provide concrete and specific data about squid populations, and in that way have sustained the reliance on fishermen begun by the ancient natural historians, while also trying to transform hazy mythic accounts into factual scientific truths. But the thought of an undersea crea-ture that is not only large enough to do battle with a whale but has ten legs and eyes like dinner plates is just too titillating to fit into a dry scientific classification. When served up as calamari, squids might seem unthreatening, but in the legends they make the humans confronting them into heroes who risk their lives in the face of strange and ferocious monsters. Beginning with Greeks in the West, squids have represented the deadliest dangers of the sea. More recent representations involve squids in sensational television tales, cult fantasies and as founders of new societies, all the while emphasizing the threat posed by alien creatures lurking in the aqueous dark. As creatures that swim in waters worldwide, squids also appear in non-Western traditions, where they play a completely different cultural role.

The primordial squid monster is the Greek Scylla, identified by Richard Ellis (an authority on sea monsters, and especially giant squids) as 'the quintessential sea monster, probably responsible for more myths, fables, fantasies, and fictions than all other marine monsters combined'.[1] In Homer's *Odyssey*, Scylla lives in a cave halfway up the sheer cliff-face of a rock. Warning her heroic lover, Odysseus, of the monster, Kirke says, 'her voice indeed is only as loud as a new-born puppy / could make, but she herself is an evil monster'. Besides her twelve feet, she has six necks, each with its own toothy head 'all full of black death'. The crafty Odysseus conceals the danger of Scylla from his men, warning the steersman only to keep clear of Charybdis – the literal gullet of the sea – and yet he arms himself, fully aware that Scylla will attack. He knows the whirlpool would kill them all, and so he willingly sacrifices the six men whom Scylla eats 'right in her doorway'.[2] Odysseus' heroism, more self-serving than that of his latter-day heirs, is signified in essential details: he wears 'glorious armour', and is

Terracotta Greek depiction of the primordial teuthic monster Scylla, 5th century BCE, now in the British Museum.

82

Scylla in a conventionalized depiction of her association with death.

especially prepared by the secret knowledge witch Kirke tells him of Scylla's inevitable attack. These details will reappear in accounts of modern heroes conquering squids.

Homer's Scylla has a high number of appendages, and more heads than any known cephalopod. But these details have never prevented fans of *A. dux* from recognizing their favourite monster. Because of the Greek anxieties regarding the dark sea – Kirke warns Odysseus that it hides 'many monsters' – accounts by sailors stress the extremity of the dangers by describing the size and hunger of the creatures. Specific details that might distinguish one monster from another get obscured by the importance of hugeness and eagerness to eat humans. Tales of the terrors lurking in the sea thus tend to be both vague and formulaic. Scylla – whom Aeschylus terms 'the nightmare of sailors' – belongs to the category of *ketos*, a term that in ancient and medieval texts can refer to multiple marine animals and monsters.[3] Only much later in the modern era does *ketos* become the specific name for cetaceans, hated foes of squids. For pre-classical Greeks it carried

a much looser meaning, including the giant squid Scylla and her cousin the sea serpent, which also holds squidly connections. The Greeks learned of these monsters from the seaworthy Phoenicians, who brought tales of them from the East to impress their trading partners with their exoticism and their bravery in crossing the seas. Goods carried from strange places on ships threatened by horrifying monsters only increased in value. The *ketos* understood to have been a cephalopod, Scylla is depicted in stone carvings most commonly as a mermaid, nakedly human from the waist up, and fishtailed below. The tail can often be multiple, but is generally singular, for the simple reason that carving multiple and sinuous appendages in stone poses technical challenges. Because of this technical difficulty, conventionalized representations of mermaids were used to refer emblematically to dreaded man-eating giant squids. Most depictions show Scylla with six heads and a dog-like body, holding small polyps with stumpy, easily carved arms. The small polyps provide another embedded emblem of Scylla's unsculptable teuthic identity, and can be seen as 'an artistic amalgam not much troubled by what swam or was thought to swim in the Mediterranean'. The canine heads – and Homer's barking puppies – refer to monstrous noises heard around her cave that could either be the cries of victims or the crashing of waves that can also destroy ships and their men. And, more directly, they remind us of her hunger for human flesh. The canine connection also works as a pun, since *skylax* is the word for puppy in the Homeric dialect. Because she lived in the dreadfully dark sea, Scylla remained nightmarishly nebulous, and was more frightening for her lack of specificity.[4]

The associations between the classical Nereid Thetis and the medieval Kraken are emphasized throughout this design of c. 1600 for a plate, engraved by Adrien Collaert.

All these qualities of Scylla – the emblematically suggestive depiction, and the uncertainty of whether she is a serpent, mermaid or squid – have fed the vast lore of sea monsters even up to the tales in our own age. Modern accounts invoke Scylla as the primordial Kraken, even the mythic relative of Aristotle's tantalizingly nebulous *teuthos*. In the thirteenth century, the German monk Albert the Great included four notable creatures in his list of 'aquatic animals'. Along with the 'luligo', which, like Pliny's 'loligo', can fly out of the water, and the inky sepia, he mentions the Beluae, 'sea-monsters' that are 'large and savage' enough to create violent upwellings that overturn ships (Alexander's navy was threatened this way). And then there is the 'Scilla', which 'often takes the form of a maiden, but has a huge gaping mouth and exceedingly sharp teeth', along with 'a craving for eating flesh'. Albert says the monster's name comes from the fact that she has the head of a dog, taking Homer's pun as a literal, but unfounded reference.[5] As one of the foremost scholars of his era, Albert relied on the major classical sources, Aristotle and Pliny, along with later redactors like Isidore of Seville to convey the important learning available in his day. The dangers of over-turned ships and of Scilla's sharp canine teeth have the same factual quality, therefore, as the 'luligo's' flight and the sepia's ink. The generic 'belua' and the recognizably emblematic Scilla become united hereafter in reports of the giant and ravenous Kraken.

Although other maritime cultures existed throughout the Mediterranean and northern Atlantic, they do not seem to have had the necessary combination of seafarers and literate historians interested in writing down their tales. So the first direct references to the dreaded Kraken appear in medieval Scandinavia. Contemporaneous with Albert the Great, the Norse text *King's Mirror* (circa 1250) offers a detail overlooked in the southern texts. The anonymous author says of the Kraken, 'There are very few

men who can tell anything definite about it, inasmuch as it is rarely seen by men.'[6] Some commentators have interpreted this comment to indicate scepticism about the factuality of the Kraken, suggesting that the author is trying to avoid 'mythology'.[7] But since the Kraken is indeed rarely seen, the statement suggests less a scepticism about the existence of Kraken and more a factual statement about the rarity of actually sighting one. In his pivotal lecture on the giant squid six centuries later, Japetus Steenstrup cites the Danish king's command to bury the beached Kraken in order to prevent dangerous talk. Since Theophrastus, the tradition has held that the sight of something as rare, large and strange-looking as a giant squid foretells some disaster, like war or the death of a prince. The indefiniteness of the descriptions has almost nothing to do with avoiding mythology, and everything to do with the fear of monsters that become more frightening in being rare and nebulous. Few can say anything definite, and few want to say anything at all, and even those should be silenced. *King's Mirror* also refers to the *hafgufa*, noting that 'it is usually called "the Kraken"'.[8] These synonyms appear interchangeably throughout the literature.

This 3rd-century BCE Greek mosaic in Calabria shows the strong association of teuthic monsters with the Ketos, namely, the horns.

In the thirteenth-century Norse saga *Arrow-odd*, the main character, Odd, and his son, Vignir, sail to the Greenland Sea to meet Odd's greatest enemy, Ogmund Eythjof's killer. On the way they encounter 'two rocks rearing up out of the sea, and Odd thought this very strange'. Farther on, during the same day, they come to an island, on which the men from one of the two ships land to look for water: 'they'd only been there a short while when the island sank and they were all drowned.'[9] When Odd and Vignir look back, they see that the rocks have also vanished. The elder Odd is 'flabbergasted', and it is up to the enlightened son to explain what they have just encountered:

These were two sea-monsters, one called Sea-Reek, and the other Heather-Back. The Sea-Reek is the biggest monster in the whole ocean. It swallows men and ships, and whales too, and anything else around. It stays underwater for days, then it puts up its mouth and nostrils, and when it does, it never stays on the surface for less than one tide. Now that sound we sailed through was the space between its two

With their large eyes, horns and serpentine bodies, the monsters in this 16th-century Dutch engraving typify the Kraken/Ketos.

The Kraken in action, from Olaus Magnus's *Historiae de gentibus septentrionalibus* (Antwerp, 1557).

jaws, and its nostrils and lower jaw were the two rocks we saw in the sea.[10]

In Old Norse, 'Sea-reek' appears as *Hafgufa*, and would seem to refer to the ammonia smell accompanying a giant squid. The accompanying monster, Heather-back, or *Lyngbakr*, appears mainly to underscore the enormity of the monsters in the sea (and of the evil power wielded by Ogmund, who will tear out Vignir's throat two pages later).

A sea creature large enough to be mistaken for an island provides one of the motifs of sea monster lore well into the nineteenth century. Sailors dock their ships, hike onto the 'island', and build a fire (or, in the case of St Brendan, hold an Easter mass) that wakens the giant who then dives underwater and drowns all the men. The arctic explorer, diplomat and scholar Fridtjof Nansen observes that this sea monster island also appears in accounts of whales (or walruses, who were called whales), and 'is doubtless derived from the east'.[11] Finding his source in the adventures of Sinbad in *A Thousand and One Nights*, Nansen makes an important

The legendary St Brendan celebrating Mass atop a kraken, mistaken for an island.

connection when he links this motif to the identical one in whale lore, reminding us that these terms hold broad reference to very large sea creatures.

Ketos, in the medieval era, held the same broad reference to sea monsters that it had for the ancient writers, and is thus similar in scope to Albert's belua.[12] Even more broadly, the term in texts like the early Christian natural history, *Physiologus*, refers to large marine creatures, and could include whales, but also giant squids and the obscure *aspidochelone*. The word 'aspidochelone' might be translated as 'shield-bearing tortoise', from *aspis* (shield) and *cholone* (turtle); though *aspis* has often been translated as 'asp' (but not in the standard Greek lexicon), so that the compound would become asp-turtle. On this basis, some Kraken historians argue that the giant 'asp-turtle' refers to the giant squid with its serpentine arms and tentacles.[13] Just as modern teuthologists rely on tally sheets kept by fishermen to record distributions of different species, so ancient and medieval natural historians relied on the

This 13th-century bestiary shows two fishermen on an *aspidochelone*, a monster synonymous with the *ketos*/kraken, and likewise often mistaken for an island.

corporis magnitudine. ac sussocatos punit.
Isa arta semitas p ꝗc Gignunt ī ethiopia.
elephantos solito more ꝗ z in yndia. ī ipso ꝼen
diuutꝰ delirescontꝰ. colla dio rugit estiſ ···
eoꝝ sub modiꝰ alligat. De belua que dicit̃ occut̃.

stories told by mariners, and then related those tales to references in learned texts. Ketos the large sea monster, Aspidochelone the large asp-turtle, and the Hafgufa or Kraken can easily stand in for one another as 'a sea monster as large as an island', which should not be witnessed by ordinary human eyes. As with Ketos, however, Hafgufa – like 'cuttle-fish', 'sepia' and 'belua' – refers less to a single creature than to the more monstrously nebulous creatures moving unseen below the sea's surface. Hafgufa and the island that drowns gullible sailors play important roles in the tradition of sea monsters encountered by the few intrepid seafarers venturing out into the always dangerous seas.

Conventions and formulaic narrative devices like these enable storytellers to frame descriptions of bizarre and frightening events in familiar ways. And this is so especially for descriptions of sea monsters that we reject too easily nowadays as fabulous or exaggerated, because modern accounts have shifted towards scientific descriptions, which adhere to their own conventions and descriptive devices. The sailors and fishermen who titillated their landlubbing listeners with stories of Kraken were not necessarily trying to delude their audience, but were relying on the descriptive and narrative devices available to them, and on long traditions of identifying strange sea creatures with real threats.

For example, the sixteenth-century writer and cartographer Olaus Magnus combined the primary texts of natural history that he found in the Vatican library, with the regional tales he had gathered from Scandinavian fishermen. He published a large, richly illustrated map of the North Sea that became the primary source for images of sea monsters. Then, twenty years later, he produced *A History of the Northern People*, in which he provided detailed accounts of people and animals, and in particular the fullest account of the Kraken in that era:

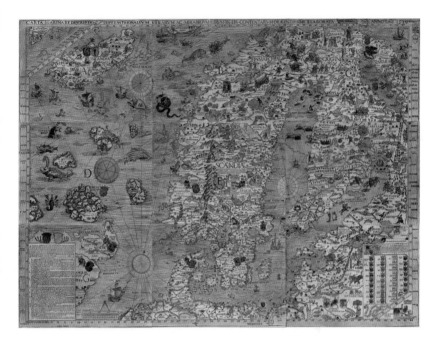

Their forms are horrible, the Heads square, all set with prickles, and they have sharp and long horns round about . . . They are ten or twelve Cubits long, very black, and with huge eyes: the compass whereof is above eight or ten Cubits: the Apple of the Eye is of one Cubit, and is red and fiery coloured, which in the dark Night appears to Fishermen afar off under waters as a burning fire.[14]

This is the famous 1539 Olaus Magnus *Carta marina* map in its entirety. Its depictions of sea monsters became the standard for subsequent images and descriptions.

Olaus lists details that had already appeared in descriptions of the Kraken for generations. The horrible head with horns or prickles sometimes appears as a horse's head with a flowing mane. The fiery eyes denote evil intent, and are occasionally displaced by a comparison to dinner plates. Just as visual and verbal

Allerley Wallfischen. CVI

Under dem Bapst Eugenio/ist bey der statt Sibinicum/in dem Jllyrischen Meer

Conrad Gessner depicts the Sea Devil, or Kraken, with the conventional flaming eyes, fangs and coiled tail.

This detail from the Olaus Magnus map shows the Kraken (on the right), with the details of convention – streaming mane, flaming eyes and whiskers (or fangs).

depictions of Scylla rely on emblematic references to puppies, polyps, mermaids, and whirlpools, medieval accounts of the Kraken repeat particular terms – which now seem outlandish – to serve as an index of accuracy. A seaman or natural historian who described a creature with these terms could certify that he had seen a kraken.

With their use of conventional motifs and terms, all these legends carried the qualities of literature. And in that way they built on the generality of reference – allowing *ketos* to name a fairly broad range of creatures – as a way of emphasizing that seafarers who ventured beyond the sight of land faced serious dangers. Following Olaus, natural historians began to look at these legends differently, asking if the conventional motifs and bizarre exaggerations might hold codified truths that could be unravelled into simple, direct references to actual – and rather ordinary – creatures. In the nineteenth century especially, commentators could denounce Dénys-Montfort as a bombastic charlatan and at the same time consider whether accounts of Kraken, and even of Scylla, might actually have been efforts to describe real creatures. Beachings of giant squids in the late seventeenth and eighteenth centuries attracted the attention of other intellectuals besides Steenstrup, and at the same time educated people throughout Europe were beginning to collect regional folk tales and to inquire into the possibility that they

Even with its modern sense of accuracy, this 1661 etching relies on traditional conventions depicting sea monsters.

might contain historical and cultural records of a sort. The second half of the nineteenth century especially saw a good number of works devoted to decoding, or modernizing, the mythic or literary aspects of sea-monster legends by eliminating exaggerations or emblematic generalities in order to uncover a history of factual references to real creatures.

Many such decoders – including Steenstrup – sought to translate the nebulosity of mythic monsters into direct references to giant squids or to other genuinely ordinary sea creatures. In 1883 Henry Lee aimed to make his mark on this modern trend by debunking superstitious tales of monsters in *Sea-Monsters Unmasked*. Without any reference to Steenstrup's work of thirty years earlier (it had not been translated into English), Lee lists the common motifs appearing in accounts of the Kraken and sea serpent (such as the head shaped like a horse, and the shaggy mane or horns), and finds that they all refer to squids. Lee explains that as

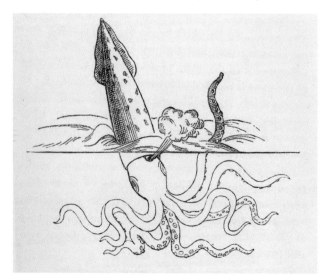

Henry Lee connected the details from legends to anatomical details of ordinary squids.

a naturalist at the Brighton Aquarium, he watched '*Loliginids*' swimming in the tanks, and 'recognised in their peculiar habit of occasionally swimming half-submerged, with uplifted caudal extremity, and trailing arms, the fact that I had before me the "sea-serpent" of many a well-authenticated anecdote'. Like Steenstrup and others since, Lee proves that monstrous legends refer to very ordinary creatures, but in terms that are at once hyperbolic and recognizably emblematic. After quoting one 'venerable' authority's account of a sea serpent attacking four men in a small boat, Lee comments, 'there is no room for the shadow of a doubt that they all recounted conscientiously that which they saw.' And then he adds this qualification: 'these perfectly credible eye-witnesses did not correctly interpret that which they witnessed.'[15] From his own experience with the aquarium squids, Lee knows the correct interpretation: the Kraken and sea serpents are all squids; large ones, to be sure, but just plain squids.

Even while reducing the monsters of legends to the squids of everyday reality, some readers of the myths still emphasize the strangeness of cephalopods. In an 1892 book, Anthony Cornelis Oudemans reviews accounts of sea serpents being seen, and repeats the warning about Kraken's portentousness, saying that 'commonly this does not happen without a terrible event in the kingdom . . . either that one of the princes will die or be banished, or that war will soon break out.'[16] In a 1918 article, J. A. Teit

Henry Lee claims this drawing, from his 1883 *Sea Monsters Unmasked*, proves that the sea-serpent is a mere calamari.

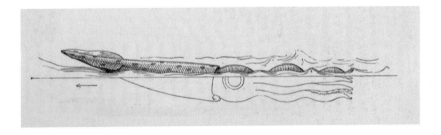

97

The *Alecton,* the ship that famously almost caught a giant squid in 1861, off Tenerife.

catalogues the 'water-beings' appearing in Shetland folklore, and includes the statement by 'an old man' about the sea serpent 'that it was not seen once in a life time, and when seen, usually portended bad luck or some calamity'.[17] The king Steenstrup refers to clearly feared for the security of his reign. Sea serpents sustain this unsettling cosmic significance of the Kraken, because they are an extension of the same legends. Oudemans admits to some doubt of the broad claims of monstrous violence against humans, stating that, 'the story of snatching away a man from the ships is evidently confounded with another tale, as it is not mentioned any where else with regard to the sea-serpent.' That other tale refers, he says, 'to gigantic calamaries which occasionally attack boats and snatch away one of the crew'.[18]

Oudemans wrote in the time of numerous giant squid sightings, which aroused considerable popular attention. Along with Steenstrup's pivotal lecture and the near-capture of a living *A. dux* by the French navy ship *Alecton* in 1861, the story of Theophilus Piccot and Moses Harvey is often recounted as affirmation that

Edward Wigfull's 1908 drawing evokes the story of Theophilus Piccot.

the monsters of myth were very ordinary squids. The 1873 event comes to read like something of a fish tale, however, in that the primary source, Reverend Harvey, delightfully reported the events at every opportunity over the years, adjusting certain details that increased the prominence of his own role. The basic story is that Piccot was fishing in a small boat in Conception Bay, Newfoundland, when a giant squid attacked. Piccot cut off two of the arms and gave them to Harvey, who had them photographed, and wrote of the event to one Alexander Murray, who in turn wrote to Professor Louis Agassiz of Cambridge, Massachusetts. Harvey popularized the event in numerous publications and speeches, in some of which two men – Piccot and Daniel Squires – were in the boat, while in others these two men were accompanied by Piccot's son, Tom, who went on to enjoy a career as a fictional character in popular children's stories.

A month after Piccot gave him the appendage, Harvey bought more squid remains from four unnamed fishermen. This carcass he photographed draped over his sponge bath to prove once and for all the existence of 'the hitherto mythical devilfish'.[19] As self-serving as the Phoenicians enticing their Greek customers, Harvey succeeded in sensationalizing a strange and violent monster – the aqueous devil – and setting himself as the gentleman scholar forsaking his bath for scientific advancement. Because he provided a photograph instead of a drawing based on a non-existent painting in a non-existent church, Harvey's tale has been taken more seriously than Dénys-Montfort's. His sensational self-promotion earned him mention in a discussion on legends, however, not of science.

Seafarers logically provide the primary accounts of devilish monsters, leaving it to the scientists, historians and documentarians to pare down or add to the sensationalism. One seafarer, Frank Bullen, from the late nineteenth century recounts sighting a giant squid directly, and reinforces the monstrosity of deep-sea devils. In *The Cruise of the 'Cachalot'* (1903) Bullen tells of being on watch one night, enjoying the reflection of 'the tropical moon' on the water, when he encountered 'the strangest sight I ever saw'. At the centre of the moon's light there arose 'a violent commotion' that made him fear a volcanic eruption. In fact, he says,

> A very large sperm whale was locked in deadly conflict with a cuttle-fish or squid, almost as large as himself, whose

The famous photograph of the Reverend Moses Harvey's *Architeuthis* specimen, draped over his Newfoundland bathtub in 1873.

interminable tentacles seemed to embrace the whole of his body. The head of the whale especially seemed a perfect net-work of writhing arms . . . By the side of the black columnar head of the whale appeared the head of the great squid, as awful an object as one could well imagine even in a fevered dream.[20]

Bullen alone witnessed a battle of giants that many have imagined. He also stands out among his fellow sea adventurers for appreciating what he saw, as he emphasizes by saying that sailors 'are the least observant of men'. Bullen's account matters not for its accuracy, especially since he almost certainly did not see any such event. Rather, like Harvey he wants to convey his own heroic (and aesthetic) capability. In reflecting on his unique experience, Bullen comments that

the imagination can hardly picture a more terrible object than one of these huge monsters brooding in the ocean depths, the gloom of his surroundings increased by the inky fluid (sepia) which he secretes in copious quantities, every cup shaped disc, of the hundreds with which the restless tentacles are furnished, ready at the slightest touch to grip whatever is near, not only by suction, but by the great claws set all round within its circle. . . . The very thought of it makes one's flesh crawl. Well did Michelet term them 'the insatiable nightmares of the sea'.[21]

The nightmare of his imagination matters as much to Bullen as having actually seen a whale eat a giant squid, for the terrible, even undepictable image of an insatiable sea monster becomes all the more troubling when it can be linked to certain reality. Bullen's imaginary event exploits – in reverse – the fashion of translating nightmares of giants craving human flesh into empirical accounts of animals we might eat.

All these efforts to give a factual grounding to myths of Scylla, Kraken and sea serpents reveal a distinctive unease in the West with the kind of knowledge and understanding provided by non-empirical modes, such as literature and legends. This unease arose after the Enlightenment, when the need for a direct observation led to subsequent requirements to arrange observations into a system of defined categories and clear relations. Certainly Aristotle and the natural historians following him observed the world around them and sought to describe the various beings

Illustration from *The Cruise of the 'Cachalot'* by Frank Bullen, depicting the event he could not have actually seen.

they encountered in a meaningful way. But natural histories worked alongside the more literary accounts of the world, judging such accounts – which could include myths, legends and folklore, as well as high literary art – as elements of a cultural landscape intertwined with the natural landscape.

A modern culture outside the West can help us to recognize that literary accounts continue to offer possibilities for engaging differently with the creatures with which we share the world. In the South Pacific people have moved over the oceans for so long that the open sea constitutes the landscape of their culture, the backdrop to their values. Polynesian cultures group different cephalopods under the same heading, not as krakenish monsters, but as sinuous creators of life and art. In Pacific cultures these animals generate a fluid interchange that, like the ocean in which they live, brings living beings together.

A Hawaiian prayer chanted to heal a sick person demonstrates some aspects of this interchange:

> O Kanaloa, god of the squid
> Here is your patient
> O squid of the deep blue sea,
> Squid that inhabits the coral reef
> Squid that squirts water from its sack!
> Here is a sick man for you to heal,
> A patient put to bed for treatment by the squid that
>   lies flat.[22]

Kanaloa, god of the squid, is called Tanagroa by the Māori. This is the Creator, from whose body all creatures emerge, and in Taluata (the Marquesas) is known as Tanoa, the god of Primeval Darkness. The Supreme Creator might be referred to indirectly as Atua, or on Hawaii as Ke Akua, which gets translated in the same prayer as either octopus or squid. The Creator's name cannot be spoken directly, however, so 'squid' provides the emblem that both conceals and manifests, depending on how we accept it.

Throughout this prayer, the term for 'squid' is *ka hee*, as in the opening line, '*E Kanaloa, ke akua ka hee*'. The reference in the

This illustration, of the *Sacramento* en route from New York to Melbourne in 1877 comes from the period in the 19th century when numerous giant squids were sighted.

final line to the squid lying flat poses a familiar referential problem to the Western reader, as squids most often move through the water or rest suspended above the floor; so the reference would appear to be to the dweller of shallow bottoms, the octopus. And, indeed, in other chants, *ka hee* and *ka hee palaha* get translated as 'octopus' and 'flattened octopus'.[23] The prayer addresses two cephalopods and Kanaloa together, for he is god, octopus and squid. These varying translations convey the fluidity of family relations in Pacific Island (Nesian) myth and culture. While the modern West uses taxonomic categories to divide and separate individuals into distinct categories, the Pacific culture can see that a god, an octopus, a squid and a man can become intertwined into an ever-transforming family.

This prayer invokes the squid as *'aumakua*, a non-human animal that serves as family guardian who can ward off threats and cure illnesses of those under its protection. Each family has its own *'aumakua*, one having a squid, while others a shark or bird. Each family will have one specific squid or shark guardian, but all the members of that species are treated as family members.[24] It is within this sort of guardianship and family relations that interchanges take place between living beings.

New Zealand poet Daren Kamali employs just such interchanges to recreate Polynesian traditions into a modern tale in *Squid Out of Water: The Evolution* (2014). The series of poems recounts the progress of Kuita, lured from his cave in the ocean depths by Manteress, who leads him to shore and changes into an island goddess, while he transforms into a man, his tentacles becoming dreadlocks. The goddess kisses him and tells him:

The Gods of Oceania have spoken
You are now a man
your name is Teuthis

Māori prow,
c. 1850, depicting
guardian creatures.

Teuthis on land
Kuita in the sea.[25]

Kamali describes his work in this book as 'poems, tales, and chants
... summoned from the islands, skies, ocean, spaces between us,
spirits, gods, cannibals, and colonizers, weaved into all types of
shapes and forms exploring all dimensions'.[26]

Through a series of sinuously interlocking episodes Kamali
weaves cultural memory – 'a squid never forgets' – into a tentacular embrace of the modern world, even as it threatens Nesian
culture.[27] Teuthis moves through modern cities from Brixton to
Reno to Dubai, while Kuita roams the sea. Each of the terrestrial
places Teuthis visits introduces him to new family members ('even
I have family in Dubai').[28] Meanwhile in the sea, Kuita is awoken
'by Lapita people / black paint on their faces / blood dripping
from their mouths'. The cannibalistic Lapita were the ancient
ancestors of the Pacific Island people, migrating from New
Guinea eastward throughout the islands sometime around 1600
BCE. Following this encounter with his ancestors, Kuita is met by
Burotukula 'swimming up from Matuku / an underwater island
most know as Pulotu'. Followed by 'A school of baby squid', the
pair 'weave the myths of Burotukula to Pulotu'.[29]

In his glossary, Kamali defines Burotukula as 'Pulotu, the
underworld'. In the broader Pacific mythic system of these tales,
Burotukula can be understood as the land of eternal joy, while
Pulotu itself can be understood as paradise.[30] In effect, then, Kuita
encounters both his ancient predecessors and future kin. Having
these myths revealed to him, he appears as Taitusi with a 'black
painted face' to chant the story of 'Tavuto the sperm whale'.
Tavuto becomes 'a beautiful yalewa', or woman, who dances
seductively to lure men into 'larst', an interchange facilitated by
'combining lust and art / in fulfillment of ancient scriptures'.[31]

The dancing of lust and art brings apparently disparate energies together into a generative interchange that also reflects the value of cannibalism as the physical transmission of energy. Bearing the painted black face of the Lapita, Taitusi chants of the sperm whale Tavuto eating her brother the giant squid as a cannibalistic dance of 'larst' fulfilling ancient wisdom. In this culture of fluid interchange, the whale and squid share the same relations to one another that humans of one family share with those of another, or that humans share with other creatures. In this regard, any creature eating another creature constitutes an act of cannibalism, a dance of 'larst'.

This dancing meal transforms predation into a weaving of scriptural prophecy, recreating ancient memory into folk tales for the modern world. The cannibalistic 'larst' works as a dance chanted through the ages,

> shaping oral traditions
> in order to be heard
> ancient contemporary folk tales
> bringing the taste of blood from the kava bowl
> back to life.[32]

These poems cannibalize ancient tales to keep the cultural tradition alive and nutritive. In chanting the tradition of the squid, Kamali transforms the ancient scriptures for the modern world, weaving land and sea creatures together into a common family. In this regard, Kamali notes, 'The giant squid is interchangeable within these pages, from squid, to octopus, to Kuita, to Teuthis (as a man) changing like the climate of the Pacific.'[33]

Throughout his chants, Kamali remembers Tagaloa, as in 'Sleep Swimming' when he recalls that 'you pounded the roots of Yaqona for Tagaloa'.[34] In Kamali's knowledge system, 'squid', and the god of creation, sing out from a deep, primeval tradition of kinship generated through interchangeability. The native girl Kamali chants of, who forgets her roots and sells 'herself to the night', reflects the modern Nesians who have lost contact with their *'aumakua* and become orphaned in the modern world.[35] Kamali chants to the *'aumakua* in order to reestablish those ties that will endure and even prosper through the cannibalizing of ancient cultures by capitalism.

At the same time that Kamali has been chanting, squids in the West continue to inspire sensational news feeds and urban legends that rejuvenate older tales of teuthic monsters. In 2012 a news story reported a 63-year-old Korean woman who found her mouth being inseminated by a squid tentacle. The headline in *The Huffington Post* was 'Squid Injects Woman's Tongue with Sperm Bag as She Eats in Korea'.[36] And the brief story states chillingly, 'it's not the first time a squid tried to fertilize a human mouth.' Not surprising, since squid coitus is essentially oral. Though a post-mortem attempt to breed with humans sounds like an urban legend, in fact the event has scientific grounding, as reported in the *Journal of Parasitology*, where readers are soberly warned of the dangers connected with eating raw seafood which might contain numerous parasites, as well as the sperm of squids.[37]

On the other hand, the account of faux calamari – of pigs' anuses being sold as frozen squid – happily has no such scientific grounding. The 2013 story had barely even been reported before it was revealed to be a hoax.[38] Both of these stories reflect an anxiety over modern industrial sources of food, and they also show that squids – as creatures people in the West consider in anything but familial terms – provide a focus for uncertain fears of a hostile world. These stories underscore the fact that we are still generating legends about squid monsters and titillating tales of heroes conquering the calamary. Many of these modern tales show up in television documentaries on sea monsters, and replicate the themes from ancient monster stories.

In the 2010 documentary *Monster Quest* from the Discovery Channel, the tale begins with a fisherman, José Raoul. This intrepid soul fishes in the Sea of Cortez, where the documentarians hope to film an *Architeuthis*. On a previous expedition, in 2006, the team was accompanied by known teuthologists Roger Hanlon and Clyde Roper. Four years later, the two academics have apparently lost their taste for monster quests, as they leave the team on their own. The narrator tells of 'legendary stories' about airmen and sailors lost at sea during the Second World War who were attacked by 'monster squid'. The divers on this quest are wise enough to take the precaution of wearing a harness attached to the boat, in case a squid tries to pull them down into the murky depths. We are reminded that these heroes will be swimming among 'animals that want to eat them' – not the legendary giant squid, but rather the 1.2-metre-long (4 ft) Humboldt. In fact, the documentary devotes the entire account of the expedition to Humboldts, forgetting the giant squid altogether. When Mr Raoul recounts the fear his fellow fishermen have for Humboldts, he tells of one man he had heard of who ventured out alone to fish for his family, and was attacked by eight Humboldt squids who pulled him down and ate him.

From that chilling account – made all the scarier by the fact that it is almost first-hand – the narrative moves to the divers talking about the proximity of hospitals, and the efficiency of air lifts, just in case. When a Humboldt is jigged from the safety of the boat, we see another Humboldt attacking it while the voice-over explains that the first squid had been 'ripped off the jig by a larger cannibal'. And when the team attach a camera to the back of a Humboldt, we get a brief glimpse of the teuthic world as the camera descends, until suddenly a squid attacks the one carrying the camera, and the view goes dark. Even though *Monster Quest* does not capture a giant squid on video, it certainly establishes Humboldts as one of the primary modern squid monsters, because of their aggression and cannibalism, which does not appear in these videos as 'larst', but only as fearsome violence. These squids appear repeatedly in television tales of marine monsters, and are quickly (and aggressively) taking their place as the modern sea monster *par excellence*.

A similar documentary, *Monster Giant Squid*, produced in 2014 by National Geographic, centres around cameraman Bob Cranston, who was attacked by a Humboldt in a previous episode caught on video and widely circulated among squid fans. To protect himself this time, he wears a ballistic dry suit with a zipper of the same sort used on space suits, and reminiscent of Odysseus' 'glorious armour'. The suggestion is again that the divers will be venturing into an utterly hostile realm to seek contact with dangerous aliens. This becomes more than a suggestion as the narrator reminds us that beneath the boat 'lies 1,000 feet [300 m] of open water and . . . sea monsters'.

And yet, Bob has put the earlier attack behind him and assures us that he wishes to interact with the Humboldts on their own terms. Towards this laudable end, he carries a plastic pipe fitted with flashing lights to simulate the squids' chromatophores. Adhering to the conventions structuring the genre of horrifying Humboldt stories, the documentary shows Bob talking with local fishermen, who warn him that the squids will eat him. Despite their dread, these same fishermen, the narrator tells us, confront the voracious Humboldts every night as they bravely harvest the '*diablos rojos*'. Bob emulates these local heroes, as he 'feels no fear, but only fascination for such alien creatures in a world so different from our own'.

As he descends into the danger zone, Bob finds himself surrounded by krill, then, 'rocketing up from below, a Humboldt squid'. Taking over the narrative himself, Bob calmly describes being 'transfixed by this enormous eye, up from the deep', as the video gives us a series of shots and counter-shots of the squid's eye and Bob's diving mask, denoting, through cinematic convention, a conversational exchange begun by the blinking pipe. As intrepidly armoured Bob watches, the squid flashes red and white, then grasps the flashing pipe with an arm, lets it go, swims away, then returns to seize the pipe again, though without any apparent hunger. Bob explains: 'The aggression is gone. It just seems interested. It circles me, as though curious. We play for a moment, then suddenly it's over.' Bob is left alone amid a few wisps of ink and memories of a fleeting friendship. When he climbs back out of the realm of sea monsters and onto the boat, Bob summarizes his experience succinctly: 'I danced with the Devil, down in the deep.'

The aggressive Humboldts are gaining more cachet, but it is still the giant squid, *Architeuthis dux*, that holds the greatest allure, not only because of its size but also because of its connection to centuries of sea legends. The most notable squid documentary, the BBC's *Legends of the Deep: Giant Squid* (2013), follows Tsunemi Kubodera's successful filming of a giant squid. While less sensational than the others, the documentary still adheres to the pattern of brave explorers descending into the abyss in search of monstrous aliens. The video opens with the titillating announcement that it will track down a 'legendary monster that has remained hidden from our cameras'. Buttressing biological field work with squid lore, the video documents the 2012 expedition to the Ogasawara Islands, east of Japan, proclaiming that the goal is 'a kind of holy grail'. A team of fifty scientists and technicians have come together for this heroic quest, armed with an underwater camera that 'has taken two years to perfect' – making it even more heroic than Bob's pipe. At the same time, we are assured that *Architeuthis* legitimately belongs in the genus of monsters when the narrator recounts numerous myths about the Kraken, and concludes that the giant squid is 'one of the earth's last great enigmas'.

Along with Kubodera, the scientific team includes Edith Widder, an expert in bioluminescence and co-founder of the Ocean Research and Conservation Association (ORCA). The team make use of a deep-sea diving vessel in order to descend to some impressive depths, between 200 and 1,000 m (650–3,300 ft). At such a remove from sunlight, Dr Widder states, 'biolumines-cence must be crucial for animals struggling to survive', for that is 'a challenging place to live'. Of course, a giant squid is not a land-based human, but a native-born inhabitant of the deeps, and so the 'challenges' really belong to the beings venturing away from the dry and pleasant land. The fact that the scientists have had to invest such monumental resources simply to enter the giant squid's realm stands as clear proof that those challenges are real and serious. Thus, Dr Widder's comments heighten the hero-ism of the modern Sir Galahad questing for the new holy grail.

After a long descent – Kubodera's first ride in a submersible – the bait suddenly becomes entwined in large tentacles. This, we are told, is the 'first ever moving image' of *Architeuthis* in the deep. Over the next few minutes, the scientists film the giant squid with its beak plainly visible, eating the bait from below. The monstrous legend changes into a portrait of beauty as the giant makes no effort to drag the scientists into the abyss, but slowly and gracefully swims around the bait, and even winks at the camera, as if to suggest that the expedition aimed at con-quering the 'last great enigma' actually created more mysteries than it resolved.

These documentaries provide magnificent footage of squids, but at the foreground lies the splendour of modern technology. In like fashion, Wendy Williams's book *Kraken: The Curious, Exciting, and Slightly Disturbing Science of Squid* (2011) perpetuates legends not of killer squids but of a handful of heroic marine biologists who study monstrous cephalopods in the name of technological advancement. Williams dramatically introduces one of her main characters braving the salty quest:

> Julie Stewart cradled her research subject in her arms. Her ponytail dripped in salt water. The back straps of her luminescent yellow waterproof Grundens were twisted tightly to better fit the slight frame of her 5' 3" [158 cm] body. She was covered in squid ink. The sun had long since set. It was mid-November, 2009, about a month away from the Winter Solstice. The sea was a bit rough. The air, a bit chilly.

A student of William Gilly at Stanford, Julie is described tagging, tracking and dissecting thousands of squids, all in the name of science, for she and the other heroes are bent on improving the lot of humanity by discovering new ways of using monster squids. The technology appears less in the devices used to approach the monsters than in the myriad benefits the scientists find that squid body parts can provide humans, including a potential cure for Alzheimer's and other neurological disorders, and new forms of camouflage for the military. 'That's why species conservation is important,' says Williams, 'not only because of conservation itself but also because those species are gold mines of possibility.'[39]

Williams delights in scenes of squid dissection. When a dead giant squid is found, scientists gather in Monterey 'to attend the gala' of dissecting the carcass. And 'Julie found herself in the middle of a scientific and media frenzy', as she bravely 'pulled on her latex gloves and stepped up to the table . . . She dug into the carcass, trying to ignore the caustic smell of ammonia.' Later we find Julie on the research boat in Monterey Bay, severing the heads of Humboldts. In her rapid-fire prose, Williams tells us that Julie slipped on the ink covering the deck, that she was tired, and that 'marine biology is not a science for the faint of heart.' In this scene, Julie cuts 'the brain stem of each squid with a knife, right below the head. She didn't seem to mind.'[40] Although she wields only a low-tech scalpel, Julie tames the dreaded Humboldts by dissecting them into the parts prized by high-tech industries.

Julie's heroism turns Monterey Bay into something of a focal point for the confrontation between cool-headed scientific investigation and the modern sensationalizing of invading hordes of aliens. The bay houses the Monterey Bay Aquarium Research Institute (MBARI), one of the foremost centres for marine biology. Beautiful Monterey has also become famous in tentacular circles for the recent incursions of large groups of Humboldts. As the narrator of yet another squid documentary – *Squid Invasion* (2010) – warns us, Humboldt squids have left their 'stronghold' in the Sea of Cortez to overtake 'coastal areas from Alaska to Chile, eating everything in their path'.[41]

The bravery of the marine biologists and videographers continues a venerable tradition of heroes – and ordinary, unscientific humans – encountering sea monsters. From ravenous Scylla to the massive Kraken, these devils have served as emblems of existential threats. While Kamali views squids as family members and purveyors of ancient cultural wisdom, mythographers like Williams and the TV documentarians perpetuate the image of squids as frightening creatures to be conquered. As sensational as these modern tales are, they work well within the tradition of teuthic legends. In fact, just as Kamali's chants keep his culture alive, these modern divers and marine biologists, armed with everything from plastic pipes to sophisticated submersibles, prove that the tradition of heroes battling squids remains very much a vital literary mode.

A full-fledged modernization of squid lore drives the 2011 novel *Kraken: An Anatomy* by English fantasy writer China Miéville. The basic plot is that the embalmed specimen of a giant squid is stolen from a natural history museum, and is held hostage as London comes very near to the apocalypse. The horror of losing the squid is felt most by the cult of Teuthex – worshippers of *Architeuthis*, and squids generally. The main character, Billy, is the staff member at the museum who preserved the specimen (or idol), and becomes the focus of the various sides of psychic warriors competing for dominance. Early on a psychic detective recounts the rise of the Teuthex cult from Steenstrup's famous article:

'The whole piece is him pooh-poohing some fairy story, and saying, "No no, there's a rational explanation, gentlemen." You could say it's where the sea monster meets . . .' He gestured around him. 'This. The *modern world*.' The stress was mockery. 'Out of fable into science. The end of an old order.[42]

Dismissal of the mythic Kraken by modern science has sent adherents of the old beliefs underground. When Billy browses through the cult's library, he encounters, alongside the Bhagavad Gita and Aztec theonomicons, 'Krakenlore. Cephalopod folklore; biology; humour; art and oceanography; cheap paperbacks and antiquarian rarities.' Once 'Krakenlore' falls from currency through scientific 'pooh-poohing', it finds its preservation in overlooked venues like 'cheap paperbacks'. Later on, one of the most ardent followers of the Kraken identifies histories of the titanic battles

The action in China Miéville's 2011 *Kraken* centres around a model like this (in the Peabody Museum of Natural History, New Haven) that has become the focus of the cult of teuthists.

as 'stories turned into pub anecdotes, in the tone of an amiable, drunken bullshitter', not unlike Bullen. Just as Kamali refuses to allow colonialism to define Nesian squid culture as debased, so the Teuthex members dive into the cultural gaps that open as modern scientists like Henry Lee deride the old tales, and there they seek to preserve a squishier mode of knowledge.

Jeff VanderMeer transforms squid lore in yet another way. *City of Saints and Madmen* (2002) is not so much a novel as a collection of narratives and treatises generated by, and sustaining the city, culture, and citizenry of Ambergris (named for the valuable substance created by whales, traditionally held to be the digested remains of giant squids). Central to the cultural life of this city is the Festival of the Freshwater Squid, which begins with a parade depicting various squid themes, and then moves into mass hangings, beatings, and dismemberment of random festival goers. The opening narrative follows the character Dradin walking in the hours before the festival, oblivious to the boys re-enacting 'the grand old King Squid sinking ships with a single lash of tentacle, puddle-bound toy boats smashed against drainpipes'.[43] The Dionysian violence of the festival re-enacts on a larger scale the sea monsters attacking and eating humans.

Except that the king of this festival is a freshwater squid, unknown, undescribed and unclassified outside the pages of this book. Along with a detailed history of Ambergris, the book includes Frederick Madnock's monograph on 'King Squid'. An Ambergrisian Steenstrup, Madnock debunks the old myths surrounding freshwater squids, and then details the exciting developments in the natural history of the freshwater squid. One notable advance is that of two 'squidologists' – Furness and Leepin – who decode teuthic 'flash communication'.[44] Using a lamp and crepe paper, Furness and Leepin project their greeting into the water: 'I AM A SQUID. HOW ARE YOU TODAY?' After a week, the squidologists discover that their subjects begin displaying letters 'on their glowing skin', which, Madnock tells us, 'hints at a higher level of squid intelligence than previously reported in even such optimistic publications as *Squid Thoughts*'. Disappointingly, the squids' message is less than inviting: 'LEAVE NOW OR I

Representation of the street images used to communicate among squid cults, as in China Miéville's novel *Kraken*.

In his *City of Saints and Madmen* (2002), Jeff VanderMeer proves that most sophisticated societies honour the squid.

WILL DEVOUR YOU, SUCK OUT YOUR MARROWS, AND USE THE BONES TO MAKE A NEST FOR MY YOUNG.'[45] As with Bob Cranston's flashing pipe, the crepe paper produces a surprising response, though not as heart-warming.

This complex apparatus does not replicate, or even gesture towards the lore of Hafgufa, Kraken or sea serpents. Instead it creates an entirely new world and culture built around squids. Just as Kamali re-enacts his culture's legends to inform modern relations, writers in the West cannibalize the old fables into modern fantasies of alternative realities. Kamali's chants and the Western fantasies energize the murky monsters science had domesticated with its definitive nomenclature. Like the Kraken, a freshwater squid has no place in the scientific taxonomy, except in Ambergris, where it serves emblematically to promote vital cultural elements. Madnock and Teuthis on land, Kuita in the sea, all perpetuate, in modern cannibalizations of ancient lore, the dynamics of squids in their variable mythic names. This dynamism reflects squids' own transformations as they move into different oceanic environments. Their endurance in all ways defines them as creatures of continual, changing movement.

4 Kinetic Squid

Squids have the potential to overthrow our traditional ways of thinking about non-humans, the way we study them, and the way we engage with them. This potential challenges us in every way. For the simple fact of squid being, *squiddity*, is movement, in the sense not only of passing from here to there, but more broadly of a dynamic potency recognized in continual change. For the ancient Greeks, the sea presented a realm of incessant and unpredictable change, and the monsters moving through the sea threatened humans with personal and cultural extinction. Squids move and change too much for us to develop the sort of domestic familiarity we have with the terrestrial animals we interact with and seek to preserve in recognizable forms like breeds. With brief lives and no inter-generational contact, they preclude any family traditions among themselves that would provide a tribal continuity, such as those among elephants, crows or whales. Among squids, each generation supplants the previous one entirely. Nor do they show any regret over seeing existing orders pass away, since brevity and change are very much the quality of their existence. The number of identified species attests to squids' tendency to move beyond stabilized taxa, compelling teuthologists to name newer and narrower genera and species in hopes of containing a dynamic fluidity. In all these ways squids disrupt our modern Western strategies for understanding animals. And it is there, in their disruptive dynamism, that we might find a lesson – challenging and unsettling as it may be – in adopting, shifting, *moving* the questions that lead us to think about, study and engage with non-humans generally.

Modern science has amassed a great deal of information enabling us to take the first steps in wondering about squids in a coherent fashion – far more so than could our ancestors in the eighteenth century. These can take us well beyond simple classifications or catch-numbers, and confront us with the challenges posed by a dynamic vitality marked by continual change that promotes anything but stable definitions. The science of squids both disconcerts and delights us, for the details of squid physiology reveal elements of a shifting, dynamic intelligence utterly alien to us in its brevity, its predatory intent and its ability to thrive in unimaginable environments. In order to consider teuthic intelligence in this way, we have to find human terms to describe a being engaging in an utterly non-human way with an incomprehensible world.

Returning to the start of Western teuthology, we recall that Aristotle constructed all his taxonomies out of the different energies he perceived among the various environments. In the simplest terms, this amounted to considering primary elements – earth, air, water, heat – as distinct yet interactive forces, and the beings inhabiting those elements as active participants shaped by and shaping the dynamic quality of the element. In this sense, the world in its basic components constituted 'a principle of movement'. The term Aristotle used (as did all Greeks) to refer to the living world was *physis*, translated by Romans like Cicero and Lucretius as *natura*, and in both cases carrying the sense of active, dynamic energy. The four elements were understood as different kinetic manifestations of *physis*, different kinds of energy analogous to the multiple gods of the ancient pantheon. Earth, water, air and fire all appeared qualitatively different, but still constituted particular kinds of energetic movement.[1]

The terms of this ancient understanding can aid our exploration of squids. In the dynamism of Aristotle's natural history, animals move as the active generation of dynamic elements. The particular element in which an animal lives – water, for example – also constitutes the animal's own physical makeup, its kinetic materiality. A squid is not simply a creature that happens to live in the ocean, as though it might as easily live somewhere else and remain essentially the same. Squids are aquatic creatures in that they are shaped by all the dynamics of the ocean, and in their particular movements within the ocean they focus and change the power of the water to accomplish their vital aims. In Aristotelean

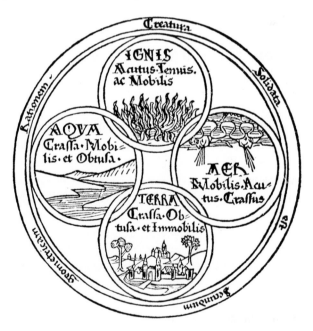

A 15th-century illustration of Empedocles' division of the four elements that Aristotle adopted for his kinetic theory to classify animals.

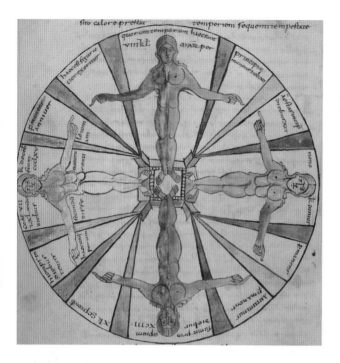

Illustration from a 9th-century manuscript of Isidore of Seville's *Traité de la nature* showing the stable organization of dynamic forces.

terms, water is the cool dynamism of fluidity that squid movements 'metabolize' into predation. Aristotle says that 'kinesis is one department of *metabolê.*'[2] For Aristotle, to metabolize something literally means to throw it beyond itself, a movement that changes one mode of being into another. Digestion of food is only one example of this dynamic process, and squids in all aspects of their being are kinetic in that they metabolize the elements through movement.

As predators – and prey – squids possess the capacity to move instantaneously and with great swiftness in virtually any direction. Like cuttlefishes, they possess undulating fins that allow them to move slowly or to hover. They also have the funnel through which they can shoot water from the cavity of their mantle, to attain jet propulsion. The very impressive squid brain contains conspicuous parallel fibres that regulate locomotion through visual cues.[3] The same parts of the brain govern both vision and movement, meaning that a squid's response to perception can be instantaneous. In fact, we might say that responsive movement and vision are simultaneous, or even synonymous, occurring not sequentially, but as the same event.

Scientific attention has focused particularly on a large star-shaped nerve ganglion in the squid nervous system which runs on either side of the mantle next to the inhalation openings. From this ganglion giant nerve axons radiate to all parts of the mantle wall. These giant axons rank among the largest among any animals, up to 1 millimetre in diameter in some species, and enable impulses to move at a much higher velocity than anything in the human nervous system, for example (they are the reason squids have attracted the attention of the pharmaceutical industry). The rapidity with which these giant axons convey information ensures that all parts of the mantle contract simultaneously – an essential requirement for jet propulsion. A squid weighing 350 g (12 oz.) can shoot out 250 ml (½ pt) of water in less than a fifth of a second. Even squids a metre or more in length can contract their mantle at the same velocity, because of the giant axon. As the mantle fills with water and then contracts to shoot it out, its circumference can change by 21 per cent and its thickness by 27 per cent.[4]

Such speed and the immediacy of metabolizing perception into attack are characteristic of predatory intelligence. The attack strategy of moving at high velocity towards the prey with arms contracted has been observed in the larvae of *Loligo opalescens* as early as four weeks, indicating that the intelligence, vigilance and agility of squids are all geared towards predation, and at a

Caribbean reef squid (*Sepioteuthis sepioidea*) showing its inescapably enormous mouth while attacking a lure.

high volume. In addition, their undulating fins combine with the jet propulsion to enable them to move through four different 'gaits', namely the hovering and manoeuvring of low speeds, and the fast and very fast jetting.[5] Some squids, mostly of the family Ommastrephidae – commonly known as flying squids – can reach enough speed to shoot out of the water and fly through the air in organized groups for distances of 15 m (50 ft) or more. In laboratories squids have traditionally caused problems by jetting out of their tanks and dying before the scientists can complete their experiments. Rebellious squids, moving with the metabolic force of their dynamic element, disrupt human control, just as water changes unpredictably.

In addition to the statocysts enabling squids to orient their movement in any direction, the lateral-line analogue – those skin cells detecting minute movements – are immediately connected

Schools of squids rely on numerous sensors to synchronize their movements; here they display their athleticism and intelligence by flying out of the water in group formation.

to axons, and serve as mechanoreceptors. Such receptors heighten their engagement in the aqueous world by enabling them to locate prey beyond the visual range as well as in low light. Squids perceive their environment visually and physically. And, because all these sensory capacities are linked directly to the ability to move swiftly, they all make for a highly successful predator.[6]

Attack strategies vary among the different species. Observations of one species of *Loligo* have shown three stages in the attack sequence – attention, positioning and strike. Accounts of the northern shortfin squid's attack offer a bit more detail. This squid averages about 28 cm (11 in.) in length, with relatively short arms. Still, it will generally attack prey its own size or larger, following the three-staged sequence. In observed attacks on trout, squids rotate from their usual tail-first swimming into a head-first position, accelerate rapidly towards the target, then capture the fish by shooting out their tentacles, and pull the prey into their open arms and snapping beak. The whole sequence takes under two seconds. This quick, predatory action combines focused speed with the terrifying flaring out of arms, a display the scientists describe as a disproportionately huge mouth overwhelming the fish.[7] During the initial phase, as the squid focuses on its victim and begins to attack, it changes colours to confuse the prey. The ability to perceive quarry, to change direction and appearance to move hydrodynamically and confusingly at high speed, and then to turn itself into a giant mouth, enables squids to attack a wide range of creatures, even those of imposing size (though probably not a sperm whale). Squids moving through a shoal of fish will very often capture one victim, only to take a few bites out of it before moving to attack another. A common reef squid has been observed to capture fifty fish in 75 minutes.[8] Unlike other predators – lions or hawks, say – squids do not gorge themselves and relax; they move to hunt and kill. And, wasteful as they are, squids kill effectively.

This continual and restless predation can be understood as both cause and effect of squids' high activity. Young squids grow fast, and therefore need to consume four times their body weight during the first nine months of life.[9] They react and move fast in

Caribbean reef squid, *Sepioteuthis sepioidea*.

order to attack a high number of quarries, in order to eat enough to react and move fast. They move, digest, change, metabolize.

As they shoot towards their victim, squids hold their arms tightly, making themselves as hydrodynamic as possible, and retract their tentacles closely. As the squid opens its arms to form the terrifyingly huge mouth, it fires out its tentacles to grab the prey and bring it back towards the flared arms and beak. Biologists have clocked acceleration during a tentacular strike as high as 250 m/s^2 in as little as 20 milliseconds, with the tentacles expanding as much as 70 per cent of their resting length.[10] The muscular structure of these appendages necessarily differs from that of the arms to allow such elasticity and speed, and indeed they are constructed of cross-striated cells, unlike the muscle fibres elsewhere on the squid's body which consist of obliquely striated cells. Such extreme elasticity of the tentacles explains why those of stranded giant squids can reach such astonishing lengths: upon death and decay, the pliable muscles stretch without maintaining their active and vital corresponding contraction.

The lateral line analogy also enables squids to move together. When large shoals of fish move back and forth in synchronous motion, they rely on their lateral line receptors to coordinate the group movements. Many loliginids appear to shoal for social reasons, spending their lives in groups synchronized in speed and direction, and swimming along parallel lines.[11] Scientists distinguish these synchronous and cohesive groups as 'schools', from the more haphazardly organized shoals. These schools of squids coordinate their movements through their lateral line receptors and their vision, enabling collective movement almost instantaneously.

Squid intelligence *is* movement, in this sense of moving through the water, and in other ways as well. The flashing colours and shimmering of their skin heighten the dynamics of their

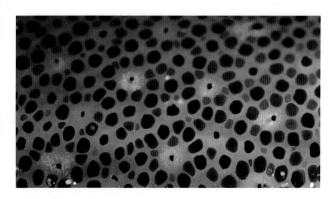

Close-up of chromatophores.

continually changing and moving arms and mantle. Anyone who has seen living cephalopods has felt compelled to remark that their brilliant colours make them among the most beautiful of all animals, attracting biologists' attention since the early nineteenth century. It was, in fact, squids still flashing on deck of the *Valdivia* that led Carl Chun to draw his beautiful hand-coloured illustrations of the various species in 1899. The actual colour changes are caused by the chromatophores, complex organs numbering in the hundreds of thousands which can be described as sacs that are opened and closed by the surrounding muscles. Muscular contraction opens individual sacs to reveal the granular pigment within, and relaxation closes the sacs to make the colour disappear. The muscles and chromatophores are controlled by the same lobes of the brain that control the funnel, making for a direct connection between jet propulsion and colouration. The nervous control of the muscles also means that expansion and contraction can take place very rapidly, with some chromatophores expanded and others closed to create patterns in the skin that are impossible in other animals. Tracing the neurological line of control, scientists have learned that chromatophores are regulated by the eyes, which send information to the regulating lobe in the brain that ensures

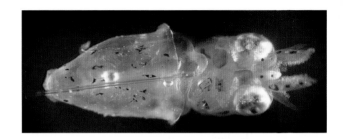

all parts of the squid's body respond together, just as in the case of the muscular contractions in jet propulsion. The neural control can also determine the brightness, contrast and colour of the patterns, which means that they can appear, disappear or change almost instantly, enabling a squid to shift its physical appearance continually to confront predators and prey with completely different shapes.[12] Not only are perception and motion intertwined, but they are also connected directly with body patterning.

Loligo pealei has three colour classes of chromatophores – yellow, red and brown – with the extremes of the chromatic spectrum marked by 'intense darkness produced by maximal expansion' at the one end, and 'intense brightness produced by maximal retraction' at the other.[13] A group of prominent biologists headed by Roger Hanlon studied the long-finned inshore squid (which they identify as *Loligo pealei*, but which Jereb and Roper identify as *Doryteuthis pealeii*[14]) for three seasons off the southern arm of Cape Cod, and in the Marine Biological Laboratory in Woods Hole, Massachusetts, observing them specifically for changes in colour, patterns created by alternate dark and bright colouring, and body shapes assumed by positioning the arms and body. These researchers sum up their observations by stating that 'any of the 34 chromatic components can be expressed instantly and in various combinations with the five postural and twelve locomotor components to produce each squid's wide variety of behavior'.[15] Examples of the components range from a 'clear' appearance (in which 'all or most chromatophores' are retracted, making a squid 'translucent in clear water or white in murky water') to 'all dark' (in which the skin takes on 'an overall deep brown coloration', and appears when a squid becomes alarmed).[16] What the researchers consider to be 'normal', because it lies in between 'clear' and 'all dark', appears as 'an overall amber body pattern'.[17]

The researchers divide body patterns into two primary types: chronic, which lasts for minutes or hours, and acute, which lasts for a few seconds or minutes. The chronic patterns all begin from the long-lasting basic amber pattern. It should not be considered the 'true' colour of these squids, however, since by their nature squids move and change. Variations depart from the amber by becoming lighter all over ('clear body pattern'), or uniformly amber along the dorsal area, while the sides become lighter ('countershading') to enable squids to blend in so well with the

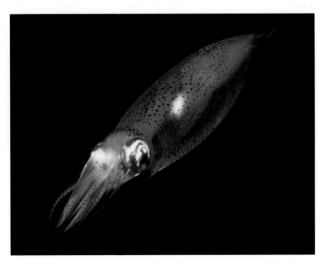

Atlantic brief squid (*Lolliguncula brevis*) flashing red chromatophores.

appearance of the water that even from a few yards from an observer they become invisible. There is also the intriguing 'chronic all dark pattern' in which an entire school of squids become consistently dark, making them stand out from their surroundings. Researchers do not understand any purpose in this shading, but have seen it repeatedly in the natural habitat when many hundreds of calm squids hover together.[18]

The brief acute patterns squids make are innumerable. Among them, 'very dark' occurs either as a single brief flash or as several flashes over the span of five seconds. The brief version appears to work as a warning of some threat coming from another squid or some other species ('a person in the laboratory'), and the longer version appears aimed at startling the other creature. Another pattern, called the 'blanch-ink-jet manoeuvre', is believed to be universal among squids. The individual blanches clear, jets away, and ejects an inky 'pseudomorph', or decoy, that remains in the approximate position from which the squid swam off.[19] This defensive action confuses an attacker that believes the ink cloud to be the squid, which has safely left the area. This 'universal' movement characterizes squids in a fundamental sense, by

Reef squid camouflaged among the coral.

These reef squids display a dark coloration, different from their more common golden sheen, both for camouflage and to communicate with each other.

creating the 'pseudomorphic' decoy that remains motionless and allows the squid to escape using jet propulsion. The effectiveness of the manoeuvre depends on its ability to convince a predator that its prey remains still rather than moving as squids do – fast and in any direction.

The chromatophores lie over reflective cells called iridophores, which change the wavelength of light bouncing off them. As chromatophores expand and contract, they cover and uncover the underlying reflective cells without extinguishing the reflected light, giving an effect of changing colours and iridescence all at once. The colours might create patterns, while the iridescence shifts the light from non-polarized to polarized, and towards the shorter wave length of blue light. The iridophores vary in thickness, which also affects the wavelengths that they each reflect. Structurally, they contain stacks of protein plates interspersed by

cytoplasm, and the thicker the protein stack, the more reflected light moves towards the longer wavelengths of yellow and red, while the thinner stacks reflect short wavelengths of blue light. This reflecting mechanism has been compared to the iridescent surface of soap bubbles, in that the tone and quality of the light shifts so continually that identification of a particular colour is impossible.[20] The angle in which an individual squid is viewed also affects the reflected light, so that seen straight on, the iridophores 'reflect red light; at oblique angles, they reflect green light and the reflected light is polarized'.[21] It also happens that the iridophores are physiologically active, and controlled by the central nervous system, meaning that they are not passive reflecting cells – like fish scales – but that squids can control the degree of polarization achieved by the light reflected from their skin. Squids can change their colour patterns (and simple colour) and iridescence according to their judgement of how they have been perceived and how they wish to be perceived, where they are in the water, and what the quality of the water is. As if this were not enough, they may also decide to appear one way to an observer in one direction and another way to an observer on the opposite side. This ability denotes an attentiveness not just to the surrounding environment but on multiple levels and directions. Like most animals with widely spaced eyes and monocular vision, squids do not force the information from both eyes into a single image, but can maintain interactions in different directions and towards opposed ends. Squids not only perceive in order to attack, they perceive how they are being perceived, and actively change the way they can be seen.

Many cephalopods, especially those spending the daylight hours at depths of 400–1,200 m (1,350–4,000 ft), flash additional light with their photophores. On squids these appear most commonly along the underside of their body and under the eyes. The control of the photophores comes from photosensitive receptors lying close to the olfactory lobe in the brain. Two groups of receptors appear to operate: one dorsally, which scientists believe registers downwelling light, and the other ventrally, which would

This close-up photograph of a bigfin reef squid shows the light colouring of the arms and the chromatophores along the eye and mantle.

register the light emitted by the ventral photophores. Generally, the photophores emit a weak blue light, matching the intensity of the light filtered through the water. On the basis of this detail, scientists surmise that the photophores on the underside function to blend individuals into the downstreaming light, so that predators from below cannot see them. By turning their laboratory lights on and off, the scientists could see the countershading strategy and decided it explained the purpose of the photophores, finding in the process as well that the squids studied could change the colour of the light they emitted to match that of their surroundings.[22]

The combination of chromatophores, iridophores and photophores (and non-ocular light perceiving organs) makes for a strongly interactive relationship with aqueous light, or rather lights, since squids engage with different types of light, and in ranges difficult for us to imagine or measure. But we can recognize that squids metabolize their aqueous environment through both water and light. Just as their squishiness, in Aristotle's

Pholidoteuthis adami (scaled squid) off the coast of Mexico. The red coloration serves as camouflage, as the blue light of aqueous depths does not reflect off red, making the squid virtually invisible to predators and prey.

terms, could be understood as water with the agency to perceive, judge and act, so their luminescence can be understood as light altered – digested, if you will – into intentional actions. Squids engage with changing, moving light as receptors, conveyors and projectors of light. They receive it, diffract it and create it. And they do so differently in different directions, at different depths and through different means. Their movements on a daily, seasonal and ontogenetic basis all participate in the environmental movements of lights, some of which we can perceive with our non-squid eyes and instruments designed to translate polarized light (as well as the light received in the non-visual photoreceptors), and much of which we cannot. The dynamics of the squid environment involve light and water interacting, and squids participate in those dynamics by metabolizing the continual changes into motion.

Squids also interact with water temperature and salinity to direct their movements. Throughout the day, many species of

A squid's eye is proportionally larger than that of any other creature, lending it a direct perception of varying types of light.

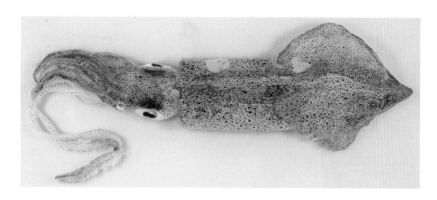

Doryteuthis pealeii (Atlantic longfin squid) showing iridophores.

squids move deeper, then return towards the surface at night. As seasonal currents change and the water grows warmer or cooler, squids move to temperatures they find more suitable to their stage in life. They may move towards the shallower waters offshore or into different parts of the ocean. And as the oceans show the effects of climate change and sustained pollution, squids move their habitual territories.

The Humboldt squid traditionally inhabits the stretch of water in the eastern Pacific from southern California to Peru, with the centre of its range located in the eastern equatorial Pacific. This notoriously aggressive squid is attracting scientific attention because of the notable increase in its range, and particularly because it has recently immigrated to Monterey Bay, home of Julie Stewart and the Monterey Bay Aquarium Research Institute. In one sense, this expansive movement has occurred as part of a periodic series of events that have been recorded since the 1830s, but in a broader sense, it strikes scientists (and sensationalizing documentarists) as unprecedented. Humboldts first began showing up in Monterey Bay in 1997 during a strong El Niño event, then disappeared until 2002 when a smaller El Niño occurred, and have stayed there since.[23] They have also appeared as far north as the Gulf of Alaska, and as far south as Chile (though in a separate population).

Along with news sensations, the appearance of the Humboldts has sparked considerable research into their migrations. Unlike the larger and more traditionally sensational giant squid, Humboldts sustain their speed over long distances. The accepted hypothesis is that they migrate northward, against the California Current, during the summer and autumn, and swim south to spawn off Mexico in early winter. The estimate that they cover something like 2,700 km (1,700 mi.) in each direction has been described as conservative. While migrating horizontally northward or southward, these squids also migrate vertically on a daily basis from up close to the surface down to depths of 350–600 m (1,200–2,000 ft), adding something like a kilometre to the distance covered daily. The southward-flowing California Current moves at an average of a tenth of a metre per second, which means that the squids,

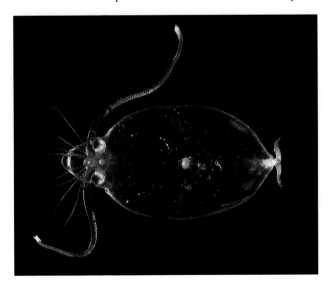

Cranchiid squid observed during the Operation Deep Scope Expedition of 2004.

which have been clocked at moving almost 40 km (25 mi.) per day, can move vigorously against the current, and will exceed the movement of the current when migrating southward. A 2012 study shows that the squids travelled at the same speed in both directions – with and against the current – and when moving south would also swim away from shore and back. In other words, while moving northward, the Humboldts kept to their intention of reaching the upper latitudes rigorously, and while swimming back south they could take time off to venture out towards deeper water.[24]

Because the Humboldt's range had formerly centred on warm waters, the dramatic expansion was thought at first to be driven by the warming of the oceans generally and more regionally during the El Niño events. But tagged squids have been traced travelling, as part of their daily vertical migration, through both warm and cool waters, through the cool California Current that flows southward along the top 500 m (1,650 ft), as well as through the warm, deeper and northward-flowing California Undercurrent, or Davidson Current. The lower current, which contains water with less oxygen, tends to create an upwelling effect as well, leading to an 'oxygen minimum zone' somewhere in the range of 100–1,000 m (350–3,500 ft) below the surface. Such zones, common to the major currents of the world, have an oxygen concentration of only around 2–4 per cent of surface waters. Although Humboldts usually stay in the oxygen-rich layers, observations have found that as climates change and the oxygen minimum zone rises, they can tolerate these depleted conditions. A 2014 study showed that Humboldts were capable of suppressing the production of adenosine triphosphate (ATP), a chemical vital to organic processes in all forms of life, but that requires continual intake of oxygen. This suppression also appears in species like the slow-moving vampire squid that inhabits the deepest parts

of the ocean. But unlike those other species, Humboldts are not characteristically sluggish, and they have a high demand for oxygen. Their capacity to endure hypoxia while retaining their predatory aggression underscores their adaptability to conditions all but deadly to most creatures.[25]

Abralia veranyi in a swarm of plankton.

These impressive discoveries about Humboldts portray an animal engaged in a wide range of motions and directionally different migrations, with the capacity to move through a low-oxygen environment to exploit environmental changes threatening many other species (including our own), and the ability to adjust its growth rate in order to exploit the other changes. These migrations occur with and against the currents, and, it is theorized, through the aid of polarized light sensitivity. Just as they metabolize the dynamics of their prey, the water and the light, squids thrive in their metabolic capacity for changeability and motility.

The major alterations threatening the existence of many creatures enrich the kinetic lives of squids. Appearances of Humboldts in new territories should be recognized as bellwether events signalling the broad changes in currents, temperatures and populations of other animals. The widening deoxygenation of the Pacific that looks to make the ocean uninhabitable for many forms of life offers new opportunities for Humboldts, whose capacity for rapid and aggressive metabolizing – or adaptability – enables it continually to move into new areas, and even those that should be deadly.[26]

Humboldts belong to a group of four ommastrephid species around the world that have expanded and contracted rapidly at various times over the past four decades. The others are *Illex illecebrosus* in the northwest Atlantic, *Todarodes pacificus* in the western Pacific, and *Todarades saggitatus* in the Norwegian fjords.[27] Because the Humboldt's particular expansion has taken place in proximity to one of the renowned oceanographic institutes, and because this squid stands out for its size, aggression and capacity to breed in such large numbers, it has attracted most attention. Whether or not the knowledge learned from studying the Humboldt applies to the other sudden expansions, and whether much at all can be applied to other squid species is uncertain. But the attention paid to the Humboldts has shown that they are even more prominent in the oceanic eco-system than previously realized, and much of their prominence stems from their expansive fluidity.

In addition to moving fast and over impressive distances, Humboldts also express their high energy in the rapidity of their growth, which can run to about 8 cm (3 in.) a month during their paralarval youth, and about 6 cm (2½ in.) a month thereafter, making them the fastest-growing species among all cephalopods. They can reach a common mantle length of 1.2 m (4 ft) in their brief life of eighteen months.[28] Scientists direct their interest in the Humboldt's characteristically rapid growth rate, mobility and high-energy hunting towards the question of how these movements might reflect changes in the dynamic environment with which the Humboldts interact. Based on Humboldts' high energy demand and the fact that available food has to change rapidly in quality and size during the squids' rapid growth cycles, scientists hypothesize that migration routes adapt to the changing occurrence of prey, and to the changing nutritional needs of the squids.[29] The view is that in both the horizontal migrations, with shoals travelling 40 km (25 mi.) a day, and the daily vertical migrations of dives down to 1,200 m (4,000 ft), the squids continually encounter different types of water mass. The various layers and gyres all have differing temperatures and kinds of nutritional opportunities (and in some cases, as in the oxygen minimum zone, serious challenges). A research group headed by Friedman Keyl attests that traits like shifts in mature size, cannibalism and range expansion give Humboldts the ability to respond quickly to the highly variable habitat of the eastern upwelling systems. According to this theory, during times (and in areas along the migratory route) when prey appears in lower numbers, due to warm waters for example, Humboldts may mature at a smaller than usual size, enabling them to expend less energy. Conversely when prey is more abundant, they mature at a larger size, enabling them to attack larger prey.[30] The El Niño/La Niña events of 1996–8 suppressed the populations of many other marine animals, creating an opening that the Humboldts exploited. At other times the Humboldt population has shrunk, upsetting the fishing industry that had grown dependent on large catches.

While the Humboldt stands out for its size and aggression, all squids are predators, and there are certainly spectacularly (and legendarily) larger species, as well as species that migrate along

The opalescent inshore squid (*Loligo opalescens*) is one of several species known for their light displays close to shore.

ocean currents and through greater depths. But its dramatic expansion and contractions in population, combined with other teuthic qualities make its kinesis notable. And certainly squids eat squids. Among the voracious Humboldts, this behaviour has been explained as part of their aggressive predatory nature, as shown in the catches containing balls of squids eating each other. During migrations, many teutologists have surmised, larger squids will eat their smaller shoal-mates in order to keep up their energy. Even more pointedly, female squids are thought to eat their male consorts in order to sustain the energy to develop their gonads, which go through most of their growth during the migration.[31] In this view, cannibalism acquires a function at the level of general population, to metabolize the species along the migratory route and through the generational and seasonal lifespan. Cannibalism within schools also demonstrates a collective behaviour that seriously challenges our notion of group identity. Recalling Daren Kamali's presentation of cannibalism as an active interchange (see Chapter Three) can help us look outside our notions of human communities and see that squid kinetics transform group identity just as they do everything else.

Other squid species, such as Caribbean reef squids, regularly form schools when they feel threatened by approaching predators. It was once speculated that reef squids set up sentinels throughout their school, with the different individuals assigned this duty facing in different directions; this speculation has since been disproved. Unlike cuttlefish or octopus species, most species of squids form shoals, though only a few come together into the closer schools. Squid groups rely on the linear analogue to respond instantaneously and collectively to movements within and beyond their group boundaries. While swimming together, squids align themselves to swim along parallel lines, and they form groups based on size. But, say Hanlon and Messenger,

> there are no known forms of grooming or play behaviour, nor is there cooperative hunting for prey, or intricate long-lasting social relationships such as those found among primates or colonial insects. Furthermore, there appear to be no complex social organizations in which there is a

Caribbean reef squids displaying light coloration with some chromatophores.

clear division of labour among individuals for activities, such as feeding, defence, home building, or care of the young.[32]

Setting aside the scientific jokes in listing squids' asocial groupings, the statements present some important qualities to keep in mind. The asociality of squid life challenges our complacent belief that a high intelligence equates to a mode of engaging with other beings and the world analogous to human socialization. We admire primates and bees because we can recognize qualities that we value in ourselves. When we see lions playing or grooming one another, our hearts melt at the assurance that the heteronormative family structure has its grounding in the natural world. But if we want to understand squids' predatory intelligence, we must confront an utterly different, even antithetical kind of community.

Social fidelity among squids probably exists for days rather than weeks or months. And it seems unlikely that aggressive, cannibalistic predators would commonly groom one another. Furthermore, for squids the act of love can lead to no promises of an eternal devotion, but only to their flesh becoming gelatinous and both partners dying. Mother squids have been seen cradling eggs in their arms, but they will never delight their babies with tales of their own youth or future dreams. Generations have almost no overlap, so that nothing like communal memory or common goals can develop. Squids grow fast, eat voraciously, swim long distances, have sex once, and die, all within the span of a year or eighteen months.[33] Large numbers of their populations get eaten by birds, seals and each other. As a metabolizing dynamism, squids flow discontinuously, adapting, migrating, invading new territory. The aggression they show as predators also makes them more than willing to take over ecological niches vacated by other

creatures. Their kinetic intelligence and brief generational spans enable them to adapt to conditions that push other creatures into extinction. Such dynamic changeability makes it hard for us to decide if we should consider squids as a single class of animals, as different creatures separated among genera and species, as shoaling generations, or as individuals engaged in their particular way with their world. Their intelligence, like the shifting colour patterns, seems to work in contrary directions, defying efforts to use our own intelligence as a measure.

Given their motile intelligence, and given that squids have shown themselves to communicate, we must ask what they communicate to one another, even allowing that we can never translate squid talk (or 'squiddish') into human talk.[34] With their squishily aggressive bodies, changing colours and light-altering flashes, do squids communicate anything more than 'I will devour you'? If

Caribbean reef squid showing its splendid array of colours.

squid shoals, and even schools, work otherwise than in a socially cohesive fashion, what desires or intentions do individuals within the groups convey to one another? Can communication be thought of as a metabolizing movement, more of change than exchange, a changing that responds to and affects the aqueous and luminescent world in which they move? Can cannibalism constitute a kind of communicability, one perhaps compelling us to consider interactions among individuals less in terms of the exchange of data and more as a metabolizing (or, to use Kamali's term, interchange) of kinetic energy?

Cannibalism within an otherwise cohesive, even cooperative group reflects the mechanical view of squids as a protein pump sustaining the biomass transfer.[35] It also challenges us to broaden our thinking of squids' predation in terms of metabolization. Squids move through water, light and populations of other

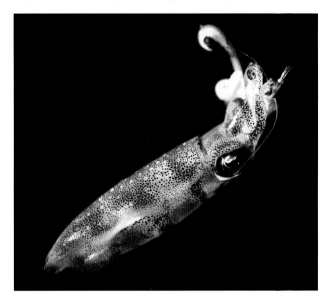

Squids communicate with a large vocabulary of colorations and body postures. Here is an example of the 'upward curl'.

Histioteuthis reversa, the reverse jewel squid. So called because its body is studded with photophores that look like small jewels.

creatures. They manipulate their 'hydroskeleton' to draw water into the cavity of their mantle and expel it.[36] They manipulate their various photosensors and projectors to receive and emit light. They exploit climatic shifts that threaten other creatures, and they eat each other. All their movements have effects, changing what they move through and what moves through them. Squids pump protein biomass from certain parts of the ocean to others, and they pump more besides. They are not merely a mechanism for transfer or exchange, however, for they embody the continual and broadest action of aqueous change – or metabolizing communication – that can extend beyond the exchange of components within the biological system. Sudden appearances of Humboldts and other squid species could signal global changes of a magnitude that can only trouble the human imagination, causing us to look at them in the way previous centuries looked at the Kraken – as portents of an anxious future.

5 Anxiety

In a taped diary entry from his effort to defraud a 1968 round-the-world solo yacht race, Donald Crowhurst recorded his thoughts on spending nine months alone in a small yacht on the Atlantic:

> You lie in your bunk at night . . . and there's a deep sub-liminal fear that something unknown [is] lurking about waiting for this moment to make its presence known – the Kraken! Unheard of terrible things that lurk in the depths waiting for Crowhurst and his trimaran! Of course, you laugh. Ha, ha you say. Ha, ha, ha. But it doesn't dispel that dark, uneasy feeling in the pit of your soul.[1]

Crowhurst at this point was descending into madness, but his rant – and perhaps even the madness itself – expresses an anxiety of which squids have become emblematic. The anxiety, as the recording shows, stems from the fact that mariners know the ocean is deep and filled with voracious monsters. Crowhurst, like plenty of seafarers before, feared that he could easily be some-one's prey.

Like the folktales and legends of the Northern seas, Crowhurst's recording reflects the anxieties of men confined to small boats in a large, violent, man-swallowing ocean. Such an anxiety – even downright terror – has spread to encompass more nebulous and

The threat that always lurks below the water's surface – from Alfred Goldsborough Mayor's 1905 *Invertebrates of the New York Coast.*

pervasive phobias than drowning. The image of a giant squid lurking below expands the sea-induced fear into broader anxieties over invasions from hostile realms. As Peter Godfrey-Smith says, confronting the nature of alien intelligence, 'cephalopods bring us into contact with something very different. . . . If we want to understand *other* minds, the minds of cephalopods are the most other of all.'[2]

Squids' status as that 'most other of all' developed out of the folklore and legends of the Kraken and sea serpents, and became fortified by scientific taxonomy to make real, modern squids into the embodiment of dread. The extreme otherness of squids gets reinforced by the modern drive to draw boundaries, such as those between species and spaces or elements. Oceans provide the human imagination with an absolute boundary as the extreme opposite of terrestrial life, teeming as they do with creatures while barring human existence utterly. The primacy Western culture awards itself has had the consequence of inducing a paranoid certainty that indeterminate 'others' are waiting to overturn that primacy. And squids, with their inhuman shape, have become the figure par excellence of a lurking, uncertain threat, because

Even though sailors and natural historians knew of calamaries, they could not imagine that the giant version could be anything other than the apocalyptic Kraken.

154

'the most other of all' swims at the greatest distance from human primacy, making a tentacle reaching above the surface appear like a warning of invading swarms.

Squids of the giant, aggressive or nebulous varieties have acquired status as emblems of anxiety so easily that they seem the natural fit. As early scientists began to establish that legends of the Kraken had some basis in fact, the possibilities of giant squids gained in their power to arouse primal fears of monsters hiding in dark places. Despite the best efforts of level-headed authorities like Japetus Steenstrup and Henry Lee, an enormous, hungry kraken could never seem as ordinary as a small market squid (*Doryteuthis opalescens*). As the last two centuries have given us the modern world of clear facts and useful knowledge, a vague anxiety persists as a possibility that we cannot laugh away. And that possibility is that somewhere in the deepest, darkest, coldest waters something large and predatory is waiting to make itself known.

Dénys-Montfort's sensationalist *Histoire naturelle générale et particulière des mollusques* (1802) had an important effect on nineteenth-century readers made nervous by political and scientific revolutions. By the second decade, the idea – uncertain as it still was – of the Kraken was common enough to appear in the jumble of topics John Keats recounted hearing the elder poet Samuel Taylor Coleridge touch upon during their walk together one Sunday in the spring of 1819: 'In those two Miles he broached a thousand things – let me see if I can give you a list – Nightingales, Poetry – on Poetical sensation . . . – Monsters – the Kraken – Mermaids – Southey believes in them.'[3] The fact that Kraken and mermaids appear in the same Coleridgean phrase together with Southey's belief, indicates that such 'monsters' had become a common topic of conversation among intellectuals.[4]

A year after Keats recorded Coleridge's incoherent ramble, Sir Walter Scott published *The Pirate*, a historical romance set in the seventeenth-century Orkney Islands. The plot of love lost is intertwined with Icelandic legends and demonology. So, when the dashing young hero is wandering along the rocky shore, the narrator offers this description of the setting:

> The Ocean also had its mysteries . . . The mermaid was still seen to glide along the waters by the moonlight . . . The Kraken, the hugest of living things, was still supposed to cumber the recesses of the Northern Ocean; and often, when some fog-bank covered the sea at a distance, the eye of the experienced boatman saw the horns of the monstrous leviathan . . . The sea-snake was also known, which . . . stretches to the skies his enormous neck, covered with a mane like that of a war horse, and with its broad glittering eyes, raised mast-head high, looks out, as it seems, for plunder or for victims.[5]

The catalogue should sound familiar, for Scott takes its key elements from the Northern legends recorded by Pontoppidan and Olaus Magnus, both of whom he had quoted in 1802 to clarify John Leyden's poem 'The Mermaid', published in *Minstrelsy of the Scottish Border*.[6] The controversy surrounding the fabled sea monsters at this juncture reflects many of the questions permeating natural history, which had to grapple with unfamiliar creatures encountered in the colonial enterprise. The strangeness of beasts and cultures made Europeans secretly fear their own world order might not be absolute.

In the 1920s, Harvard professor of English John Livingston Lowes traced Coleridge's reading through great numbers of works leading up to the ballad 'The Rime of the Ancient Mariner' (1798). And it is worth recalling two stanzas from that poem:

Beyond the shadow of the ship,
I watched the water-snakes:
They moved in tracks of shining white,
And when they reared, the elfish light
Fell off in hoary flakes.

Within the shadow of the ship
I watched their rich attire:
Blue, glossy green, and velvet black,
They coiled and swam; and every track
Was a flash of golden fire.[7]

The Mariner has committed some vague crime that turns all of nature against him – or, at least all of oceanic nature. A ghost ship sails by, and then the Mariner's own mates die, while impossibly continuing to sail the ship. Everything the Mariner encounters is impossible, or – like the water snakes – creepy. Coleridge knew virtually nothing of ships and the sea, but he did read voraciously of voyages to the antipodes, and he obviously digested their descriptions of sea monsters and other terrors. Walking with Keats, twenty years after writing his nightmare poem, Coleridge still saw the sea and its creatures as reflections of his vague anxieties. In discussing the wide-ranging sources for Coleridge's imagery, Lowes considers the modern reception of earlier accounts of strange animals, observing that 'monsters were the first to fade into the light of common day', as 'chimeras at close quarters merge into the tangible forms of which they were projections'.[8] The monsters of yore might have faded, but the nightmares did not.

A generation later, the young Alfred Tennyson wrote the poem known to virtually all latter-day teuthists, getting right to the essence of the modern anxiety represented by the Kraken. Even

'Beyond the shadow of the ship, I watched the water-snakes.' Gustave Doré's illustration to Coleridge's 'Rime of the Ancient Mariner'.

more than Coleridge, Tennyson found the major intellectual and cultural advances of the day upsetting. Particularly troubled by incomprehensible spans of geologic time and the fact of species extinctions, he situated the Kraken 'Far, far beneath in the abysmal sea'.[9] As yet unphotographed and unclassified, the Kraken could be dreaded on its own or as the sea serpent, either of which had been proclaimed by the science of the day as one of the worst dangers in the ocean deeps. Beyond everything else, the Kraken lies in 'His ancient, dreamless, uninvaded sleep', and 'There he has lain for ages.' This primordial being is so alien that his sleep – which is the only mode in which he has ever existed – is empty. He does not dream of us or his companions, because he lies so removed from our world that there is nothing for him to dream of. At some distance from him,

> far away into the sickly light
> From many a wondrous grot and secret cell
> Unnumbered and enormous polypi
> Winnow with giant arms the slumbering green.

Like the Ancient Mariner's 'water snakes' that live on as 'a million million slimy things', these polypi threaten to overwhelm the imagination in being 'unnumbered'; their quantity makes them indeterminable as specific beings, and without specificity they become repellent, creatures we turn from and avoid even thinking about. Whether these enormous beings are octopuses or squids is irrelevant at this cultural juncture when divisions among cephalopods had still not settled definitively onto the specific categories we use today. The fluidity of the terms adds to the power of the polypi to evoke disgust, but not of a sort that can be ended through taxonomic catharsis: the 'slimy things' and 'giant arms' winnow through the cultural imagination to sustain a sense of dread by evading nameable specificity.

For all their creepily unnumbered indeterminacy, these polypi in 'sickly light' and 'slumbering green' provide only a foretaste of the more fearful sleeper, the Kraken. Christopher Ricks says this poem 'is quite other than a science-fiction or Loch Ness fantasy; its depth of feeling comes from Tennyson's pained fascination with the thought of a life which somehow is not life at all'.[10] In

focusing on a hidden and nebulous threat, a life that is not life, Tennyson confronts his 'pained fascination' with the possibility that some force could overturn existence, and the possibility of human extinction. His Kraken spans the full range of life on Earth, as its millennial age precedes every other life form, and its death marks the apocalypse, when 'the latter fire shall heat the deep; / Then once by man and angels to be seen, / In roaring he shall rise and on the surface die.' It takes the world-ending fire to rouse the

The lurid sensationalism of Shannon MacGregor's 2008 mixed-media *Kraken* corresponds to that of apocalyptic literature.

William Blake's 1805 watercolour *The Number of the Beast is 666*, a visionary representation of the unrepresent-able beast related to Tennyson's Kraken.

sleeping Kraken, whose waking life consists of the single act of a roaring rise to death – its own and the world's.

Tennyson's evocation of the apocalyptic force lying in wait has remained a powerful motif in literature ever since. There is the sense of an untapped potency that humans might dream of harnessing and exploiting; but this dream is haunted by the concern that the potency has slumbered for a reason long forgotten, and to awaken it would put the world in peril. Here its awakening portends more than its own death. The connection between Tennyson's poem and the image from Revelation 13:1 of the beast rising out of the sea makes the Kraken's awakening into death the direct sign of apocalypse. Kraken death will henceforth mean world death.

Indeed, in the year after Tennyson published 'The Kraken' in *Poems, Chiefly Lyrical* (1831), his American admirer, Edgar Allan Poe, paid homage in an early work titled 'MS Found in a Bottle'. In this story the narrator, who assures us that 'no person could be less liable than myself to be led away from the severe precincts of truth by the *ignes fatui* of superstition', sails into the South Pacific, where he is overwhelmed by a powerful storm.[11] For six days the ship is caught in the darkness of 'eternal night': 'All around were horror, and thick gloom, and a black sweltering desert of ebony'. Swept into the dreaded Southern Seas, the narrator feels 'the utter hopelessness of hope itself', as he prepares himself for certain death. 'At times', he says, 'we gasped for breath at an elevation beyond the albatross – at times became dizzy with the velocity of our descent into some watery hell, where the air grew stagnant, and no sound disturbed the slumbers of the Kraken.'[12] The reference to the albatross is not accidental, as it brings Coleridge's poem of dread together with Tennyson's beast.

The portent of the sleeping Kraken transforms the certain death of the narrator into something much worse, the 'life which somehow is not life at all'. Since the portent is combined with a

Arthur Rackham's illustration for Poe's 'MS Found in a Bottle', where the ghost ship plummets to the oceanic depths where the Kraken waits.

reminder of that other strange portent, the albatross which the Ancient Mariner killed, it is logical that Poe's narrator escapes death by finding himself on a ship worked by ghostly sailors very like the undead on the Mariner's ship. In the captain's face 'the thrilling evidence of old age so utter, so extreme . . . excites within my spirit a sense – a sentiment ineffable'. Like the Mariner, this

163

Ancient Captain induces something troubling in those who look upon him or listen to him. And like Tennyson's Kraken, the troubling ineffability holds a connection to an apocalyptic agedness. Confronting these beings means confronting the world at its temporal and spatial limits – beginning in the primordial deep and ending in our encounter with that deep. The ghost ship sails into gigantic southern storms, leading the narrator to believe he is 'doomed to hover continually upon the brink of eternity'. As the icebergs close in, the narrator realizes the ship has gotten caught in a 'howling and shrieking' current moving them 'like the headlong dashing of a cataract'.[13]

Folklore often sets the Kraken alongside deadly whirlpools, at least since Homer's pairing of Scylla with Charybdis. In his *Natural History of Norway* (1755), Pontoppidan cites a scholarly speculation of his day that Odysseus encountered the teuthic Scylla and the Charybdian vortex off the Norwegian coast, where the powerful current known as the Moscoström – commonly referred to in our time as the maelstrom – creates just such nightmarish whirlpools. The island of Skarsholm and the whirlpool

Olaus Magnus' 16th-century depiction of the maelstrom derives from the ancient relation between Scylla and Charybdis.

Vanders Skyllen, so the idea goes, gave rise to the names Homer uses.[14] Whether or not Poe knew of Pontoppidan, he clearly understood the apocalyptic significance of the Kraken, and that the watery cataract adds to the inevitability of the end. Quite simply, his narrator confronts complete destruction, ending his narrative with the agonizing and disconnected phrases, 'we are plunging madly within the grasp of the whirlpool – and amid a roaring, and bellowing, and thundering of ocean and tempest, the ship is quivering – oh God! and – going down!'[15] The 'roaring' once again evoking the Kraken of apocalypse, the narrator describes more than his own death, as he has been driven to the cataract by the crew of death-in-life sailors who have shown him 'some exciting knowledge – some never-to-be-imparted secret, whose attainment is destruction'. Just as Tennyson's Kraken awakes to die roaring in the end-of-the-world fire, Poe's narrator finds his revelation in complete universal destruction.

Poe's story links the Kraken, the vortex, and annihilation of those encountering them into an apocalyptic revelation. Two motifs associated with the Kraken from folklore are that its descent creates a whirlpool capable of pulling boats down with it, and that its appearance portends some disaster, that 'princes will die or will be banished, or that a war will soon break out', or even worse.[16] With these motifs in mind, we can see that the Kraken of modern literature embodies an anxiety going beyond personal destruction to full-scale apocalypse.

Two decades after Poe's narrator plunged into the abyss, one of the most apocalyptic novels ever written was published. Herman Melville's *Moby-Dick* (1851) adumbrates the final disaster with a teuthic omen. On a 'transparent blue morning, when a stillness almost preternatural spread over the sea', the crew of *Pequod* watch in expectation as 'a great white mass . . . rising higher and higher, and disentangling itself from the azure, at last gleamed before our

Harry Clarke's
illustration for
Edgar Allan Poe's
'Descent into the
Maelstrom'.

Harry Clarke's
illustration for
Edgar Allan Poe's
'Descent into the
Maelstrom'.

prow like a snow-slide'. The slowness of the white mass troubles
the watchman: 'It seemed not a whale; and yet is this Moby Dick?
thought Daggoo.' As the boats row out in hopes of confronting
Ahab's nemesis, the mass rises and sinks repeatedly:

A vast pulpy mass, furlongs in length and breadth, of a glancing cream-color, lay floating on the water, innumerable long arms radiating from its centre, and curling and twisting like a nest of anacondas, as if blindly to clutch at any hapless object within reach. No perceptible face or front did it have; no conceivable token of either sensation or instinct; but undulated there on the billows, an unearthly, formless, chance-like apparition of life.[17]

This description holds little of the overwrought horror evoked by Poe's vision of the Kraken. Yet the 'innumerable' arms moving 'like a nest of anacondas' and the facelessness reflect the dread of *Pequod*'s crew driven by yet another maniacal captain. The realism of the description, no less than the legends of flaming red eyes, makes this squid embody pointless malevolence. It moves its arms 'as if blindly to clutch at any hapless object', and it lacks both sensation and instinct. In Ishmael's estimation, it is barely even an animal.

The final clause of this description emphasizes the depth of Ishmael's anxiety. As a formless 'apparition of life', the pseudo-animal sustains the horror of the life that is not life evoked by Tennyson's apocalyptic Kraken, and of the undead crews dreamt up by Coleridge and Poe. Being 'chance-like', it has no meaningful place in nature, but only looms without meaning, place or form. In that regard, it can only be understood as 'unearthly', or disgusting. When seaman Flask asks what the strange being is, Starbuck answers, 'The great live squid, which, they say, few whale-ships ever beheld, and returned to their ports to tell of it.' To this warning, Ishmael provides the gloss, 'certain it is that a glimpse of it being so very unusual, that circumstance has gone far to invest it with portentousness.' The rareness of the giant squid aligns it with the rare white leviathan (just as both have been linked together since medieval legends), and the appearance of the whale will verify Starbuck's portent. The *Pequod* meets its end when Moby-Dick rams it with 'his predestinating head', and it goes down in apocalyptic style, spinning with 'all its crew, and each floating oar, and every lance-pole, and spinning, animate and inanimate, all round and round in one vortex'. In the final moments before the tops go under, a hawk meets its end as Tashtego nails its wing to the sinking mast: 'and so the bird of heaven, with archangelic shrieks . . . went down with his ship, which, like Satan, would not sink to hell till she had dragged a living part of heaven along with her'.[18] Maybe this passage does get a shade purple, but the grandiosity is also appropriate to the fulfilment of the Kraken's portent of 'predestinated' disaster by way of whirlpool.

Moby-Dick completes the transition in squid depictions from tentative legends and rumours to definite and certain identifications. The squid appears as itself, uncloaked from the mythic questions of whether such enormous cephalopods exist, whether they are actually squids, and whether they are the leviathan or horned serpent or even something worse. The purpleness of Melville's prose upholds the apocalyptic extreme of the events narrated, but the monsters are creatures that could be encountered by real seafarers, even if only once in several lifetimes, namely a white whale and a giant squid. When Melville wrote this novel, modern science had still not granted its imprimatur to *Architeuthis*, but to name a creature 'squid', even while describing it as an accident of nature and as a mere blindly reactive zoophytical force, brings it into being as a reality rather than a metaphor of outmoded beliefs. And its verification continues to portend disaster.

Fifteen years after Melville's apocalypse, Victor Hugo sustained the tradition of the Kraken. His villain in *Toilers of the Sea* is seized by a monster and drowned. The hero is similarly attacked, and in the same spot, but fights free. The creature itself is given a

Jean-Henri (Henry) Dunant, *Diagram of the Apocalypse*, 1890, mixed media. This map, drawn at the time when numerous sightings of the Kraken convinced many people that the apocalypse was immanent, shows precisely how human extinction will occur.

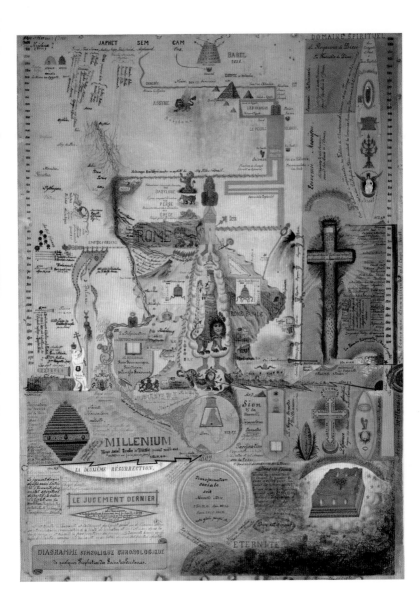

variety of names: 'This monster is termed by sailors a "poulp," the scientific name of which is cephaloptera, whilst in legendary lore they are known as Krakens. English mariners call them devil-fish and blood-suckers. In the Channel Islands they are spoken of as pieuvre.'[19] Hugo's terms, 'poulp' and 'pieuvre', are both commonly translated as 'octopus' in later English editions of the novel. Indeed, the descriptions in this story give the monster a mere eight arms. But the Kraken has also been identified as *poulpe*, and the fluidity of terms and appendages that has accompanied squids since Homer and Aristotle certainly persists here. Hugo also says that when his monster swims 'it resembles a sleeve with the closed fist in it, sewn up at the cuff',[20] in which we find a reminder that in the *Hallieutica*, Oppian also termed his squid a 'sleve'.

Hugo needs some blindly malevolent creature to lurk in the obscurity of his sea, and is less concerned than Melville with taxonomic precision. His description offers no meaningful physical details, but Hugo goes well beyond Coleridge and Tennyson in having his sleeved fist foreshadow a hostile reality supplanting our own. 'What can be said of these treasons of creation against itself?' he asks, and answers thus:

> they exist, yet their existence is improbable. They are, though reasoning is against their existence . . . You refuse to accept the vampire, and so the pieuvre appears. Their existence is a certainty which disconcerts our certainty . . . They mark the transition of one reality into another. They seem to belong to that commencement of terrible creatures which the dreamer sees through the loophole of the night.
>
> This generating of monsters, first in the invisible and then in the possible, has been guessed at – perhaps, even

perceived – by the magi and philosophers in their severe ecstacies [*sic*]. From this comes the conjecture of the existence of a hell.[21]

Following the association of Kraken with life that is not life, Hugo makes his sleeve into the treasonous creation of hell, an entirely different, improbable reality. It is such a revolutionary world that China Miéville constructs in his fantasy of the teuthex cult 140 years later.

Despite his outrage over the *pieuvre*, Hugo wanted his novel to be a realistic depiction of actual maritime dangers. Realist fiction like his works is only a step away from science fiction, which enumerates factual details to lend an air of naturalness to outlandish plots. And, indeed, the opening pages of Jules Verne's *20,000 Leagues under the Sea* (1870) play on the sensational news of real giant squids being sighted in the sea or washed up on shore. So we are told in the second paragraph that 'For some while, ships at sea had encountered "an enormous thing," a long spindle-shaped object, sometimes phosphorescent, infinitely bigger and faster than any whale.'[22] The shape of this leviathan (like a sleeve), along with its phosphorescence, can only call to the teuthophobic mind a giant squid. Verne continues his tease by intertwining references to sightings from named ships on specific dates (made up) with repeated expressions that a 'phenomenal creature' did exist as 'an undeniable fact', that the old legends had been proved correct in their kernel.[23]

But, of course, the 'creature' turns out to be only a phenomenal submarine made out of indestructible material, manned by a not quite ghastly crew who speak in an unknown language, and commanded by the evil Captain Nemo, who takes as prisoners the narrator (Professor Arronax), his valet and amanuensis (Conseil) and a Canadian harpooner (Ned Land). Since Arronax is one of

Victor Hugo's drawing of the *pieuvre*, a nightmare creature that made Hugo believe in the existence of hell, made contemporaneously with his novel *Travailleurs de la mer* (1866), which introduced the Guernsey dialect word *pieuvre* into French.

the foremost natural historians in the world, the imprisonment is not without its delights, as he and Conseil get to see hundreds of sea creatures. Ned Land, whose pleasures in life consist mostly in killing animals, grumbles incessantly, and can only hope to escape. The conjunction in the narrative of the futuristic submarine with a hike along the sea floor to Atlantis – 'the very place where the contemporaries of earliest humanity had walked' – sets the stage for an encounter with the Kraken, itself the embodiment of primordiality and apocalypse.[24]

The significant chapter of Verne's narrative in French is 'Les Poulpes', and is translated into English as 'Giant squid'. The translators offer the comment that 'in the 1860s', when Verne wrote the novel, *poulpe* 'was simply the *generic* term for an animal having numerous feet or tentacles', and that it is 'the French equivalent of Kraken'.[25] In thus translating the chapter title, they make a direct connection to Melville's chapter and contextualize Hugo's hellish spectre. Arronax and his friends do not see the squid floating on the surface, however, but, like their direct heir, Dr Kubodera, they encounter the giant squid at the bottom of the ocean.

When they see the actual giant squid through the window of the *Nautilus*, Arronax describes it as being 8 m (27 ft) long:

It was traveling backward with great speed, right alongside us. It gaped with enormous eyes tinted sea green. Its eight arms – or more accurately, legs – were rooted in it head . . . and they writhed like the serpentine hair of the Furies . . . The monster's mouth, a horn-covered beak shaped like a parrot's, opened and closed vertically. Its tongue, also made of a hornlike substance and armed with several rows of sharp teeth, would flicker out from between those genuine shears. What a freak of Nature![26]

Not in the least put off by the gruesome appearance of the monster, Arronax proclaims that he would like to dissect a specimen. When the submarine surfaces to accommodate his scientific curiosity, its hatch is pulled off, and 'immediately one of those long arms glided like a serpent into the opening and twenty others were quivering above it.' Nemo, standing with one of his crew, cuts off

the first arm, but then 'two more tentacles lashed the air, seized the seaman in front of Captain Nemo, and carried the man off with irresistible violence.'

But it gets even worse. For as the hapless crewman is swung through the air 'at the mercy of that elephantine trunk', other

Captain Nemo uses his hatchet on the evil squid, in an illustration by Neuville and Riou for Verne's *20,000 Leagues Under the Sea*.

squids crawl up the sides of the ship, compelling the entire crew to hack at the invaders: 'A violent, musky odor filled the air. It was horrible.'[27] As the battle continues, Ned Land wields his harpoon to plunge it 'into a sea-green eye and burst it', only to be entwined in a serpentine tentacle in his own turn. Arronax expresses his dread thus: 'The squid's formidable beak was opening right over the fallen Canadian. The poor man was about to be chopped in two.' Of course, he is not, because Nemo – the only character exceeding Ned in manly violence – buries his axe in the beak, enabling Ned to plunge his harpoon 'deep into the squid's triple heart!'[28]

What stands out most in this thrilling scene (who does not love to wince at the image of the 'genuine shears' of the giant beak chop, chop, chopping over the virile Ned?) is the intertwining of modern anatomical knowledge with folkloric sensationalism. Verne was clearly writing for an audience that had made the transition from Kraken to giant squid, while retaining fears that even science cannot forestall the apocalypse portended by the desecration of Atlantis. The monstrosity of the squid attack lies in Arronax's dispassionate detailing of the creatures, their serpentine tentacles with elephantine suckers, the 'musky odor' of ammonia and the unmanning mandibles. With this one scene, Verne has succeeded in casting the giant squid as the violent and repellent embodiment of the mythic fears transformed into empirical dread.

As predators, squids do not necessarily have to appear as Kraken or giant squids to evoke uneasy feelings among humans. Wendy Williams mythologizes her scientists for their confidence that their empirical knowledge can harvest creatures for their pharmaceutical potential without inciting any existential treason. Such a myth casts human predation as the pursuit of knowledge and as the development of resources for welfare. In stark contrast to Williams's adulation of human (or corporate) primacy, William Hope Hodgson envisions another, hellish reality.

First published in 1904, *The Boats of the 'Glen Carrig'* recounts the travails of seamen drifting in the ship's boats following some unstated disaster in the year 1757. After a storm, the men find themselves trapped in a sea of weeds ancient and dense enough to have entangled several ships that have rotted into ghostly hulks. Noticing 'strange movements', the narrator, John Winterstraw begins to see, 'as one may in dreams, dim white faces' peering at him from below.[29] He soon has a direct encounter as he leans over the edge of the boat:

> I found myself looking down into a white demoniac face, human save that the mouth and nose had greatly the appearance of a beak . . . there came a sudden, hateful reek in my nostrils – foul and abominable.[30]

Winterstraw's account of the face as 'demoniac' is descriptive within the conventions of apocalyptic stories, for it is that of a squid, and the first of countless squids, some giant, others less so, but all predatory.

The men in the boat find refuge on an island, only to learn that it offers little protection from the squids, or devil-fish, who can slither up the beach, and use an underground passage to attack the men's camp from various sides. And attack they do, every night and all night. Standing watch on a promontory, Winterstraw observes one attack in preparation: 'a number of great fish were swimming across from the island, diagonally towards the great continent of weed: they were swimming in one wake, and keeping a very regular line.'[31]

A great battle ensues, as the swarms of 'unearthly' squids attack the camp. Winterstraw recounts that he had to fight his own urge to retch from disgust and horror as much as he had to fight the squids themselves: 'there came into view, not a fathom below my feet, a face like to the face which had peered up into my own on that night. . . . the great eyes, so big as crown pieces, the bill like to an inverted parrot's, and the slug-like undulating of its white and slimy body, bred in me the dumbness of one mortally stricken.'[32] Winterstraw's disgust and horror, like Hugo's, derives both from the thought of being eaten and by confronting the demons aiming to replace the order of existence with another reeking of ammonia.

Hodgson's squids display the intelligence of a disciplined and organized attack, with a clear intention, namely harvesting the humans. A powerful intelligence focused on using humans as a resource is a theme of hellish nightmares. The ambition of Professor Arronax and his scientific heirs to harvest and dissect large numbers of squids becomes a sick little joke in the face of such a large and organized onslaught. The indeterminacy of the 'swarm' of squids also reverses the view human predators perpetuate of their prey, that they are barely animals, so to harvest them causes no real harm, or pain either. Reduced to prey animals, Winterstraw and his crew confront an onslaught of predators as an overwhelmingly indeterminate threat, against whom human individuals, with their 'higher' consciousness, need to struggle to avoid being depersonalized into biomass. From here on, squid monsters carry their mythic inheritance into portentous stories of alien invasion threatening humans collectively.

The archetype is the account of a primordial squid race lying dormant, as in H. P. Lovecraft's famous story 'The Call of Cthulhu' (1926), which casts the indescribable 'Thing' as a cephalopod. At various times in the story, the bas-relief images and the creature itself are called an octopus, cuttlefish, 'squid-dragon' and finally an 'awful squid-head with writhing feelers'.[33] Lovecraft strategically sustains the terminological slippage to summon up an alien so inconceivable as to exceed a simple identification. Were Cthulhu to appear as a serpent, say, or a slug from a black lagoon, he would be creepy enough, but too recognizably terrestrial to drive mad anyone who sees him directly.

Like Tennyson's Kraken, Cthulhu has slept innumerable ages at the bottom of the sea. The chant sung by 'diabolist Esquimaux' and Louisiana swamp-priests – '*Ph'nglui mglw'nafh Cthulhu R'lyeh wgah'nagl fhtagn*' – translates into 'In his house at R'lyeh dead Cthulhu waits dreaming'.[34] Cthulhu is one of the 'Great Old ones who lived ages before there were any men, and who came to the young world out of the sky'.[35] Not only are they aliens from another distant world, but from a time so ancient as to make 'man and the world seem recent and transient indeed'.[36] As with the

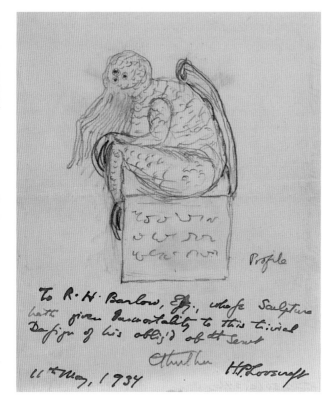

H. P. Lovecraft's original sketch of Cthulhu.

Kraken, it is important that these 'old ones' have inhabited our world aeons before it was ours, and that they have lain dormant 'in the mighty city of R'lyeh under the waters', for it is thus that they reinforce their ominous alienness.[37] They are as unlike humans as can be, and unlike anything conceivable to humans, since they come from a world in another galaxy, they ruled this world before it was ours – or 'this' – and they sleep in the part of this world that humans cannot enter. The cephalopodal designations, then, incorporate all these degrees of alienness to depict a creature that 'disconcerts our certainty'. And the most disturbing aspect of all is that the world we have claimed as our own has belonged to them all along. Intelligent beings, whose intelligence is unlike anything we could recognize as such, lurk here with us, unknown, and unknowable, but waiting, waiting.

John Wyndham's 1953 novel *The Kraken Wakes* confronts the unknowable quality of invading squids head on, and with a disconcerting prescience of apocalyptic threats to our own time. Like Cthulhu, the beings in this narrative come from outer space, and take up residence in the deepest parts of Earth's oceans. Unlike the other works that set squids at the centre of overwrought narratives about monsters and hellish revelations, Wyndham's story focuses on two journalists, the husband and wife team of Mike and Phyllis Watson, working for the English radio outlet EBC – continually confused with the *other* broadcasting network. As journalists, Mike and Phyllis spend most of their time interviewing authorities who propound their learned and dispassionate opinions on what the aliens are and how we humans should deal with them.

The brilliance of Wyndham's novel lies in its depiction of human fecklessness in the face of imminent disaster. Three years pass after the aliens first splash into the ocean, and people lose interest in them. It is only when world trade is halted by the deep-sea colonizers sinking ships that the major powers – which is to say, financiers first and the military next – decide to act by electrocuting the creatures. The strategy fails, as the aliens easily disable all human weaponry. The authorities' ignorance and violence trickle down into denialist pub rants, like this by a 'medium-built man':

> 'All right', he said, 'say for the sake of argument they're right, say there *are* these whatsits at the bottom of the sea; then what I want to know is why we're not getting after 'em right away? What do we pay for a navy for? And we've got atom bombs, haven't we? Well, why don't we go out to bomb 'em to hell before they get up to more trouble? Sitting down here and letting 'em think they can do as they like isn't going to help. Show 'em, is what I say, show 'em quick and show 'em proper. Oh, thanks; mine's a light ale.'[38]

Dr Alistair Bocker, an 'eminent geographer', whose name is 'customarily followed by several groups of initials', is the one person who expends any responsible thought to the deep-sea inhabitants. Unfortunately, he is denounced as a crackpot, and the financial backers who had funded his efforts to learn about the beings pull out for fear 'that their product's reputation would suffer by being associated' with the man.[39] The derision comes as a result of Bocker stating openly that humanity was witnessing 'a species of interplanetary invasion'. He lowers his prestige even further when he argues that more than a single species of intelligence exists, and that 'the greatest efforts should be made to develop some means of making a sympathetic approach to the new dwellers in our depths with the aim of facilitating an exchange of science.'[40] The contrary view is summed up tersely by one of the Watsons' fellow journalists: 'Can you imagine us tolerating any form of rival intelligence on earth, no matter how it got here?'[41]

After all the heavy-handed efforts to destroy them fail, the deep-sea colonizers – who have come to be known as Bathies – commence highly orchestrated attacks on humans, while also altering the ocean currents, upon which world civilization collapses. The Watsons watch London flood, and then escape by boat to their hilltop cottage in Wales, agonizing over dwindling food supplies. In the despair of seeing their world end, Phyllis reflects, 'sometimes I dream of them lying down in those deep dark valleys, and sometimes they look like monstrous squids . . . there they are all the time, thinking and plotting what they can do to finish us right off so that everything will be theirs.'[42] Phyllis's dread reminds us what it means for the Kraken to wake – the treasonous creation of hell, the apocalyptic end. And it reminds us that squids, by the simple fact of their existence, and of their existence in environments that seem impossible to humans, remain the epitome of beings as unlike humans as possible. That such a creature should exist is troubling enough, but that it should possess the high intelligence of a predator makes it into an enduring emblem of human anxiety over the possibility of inconceivable world orders replacing our own. The image of creatures preying on humans as humans prey on them is one that we turn away from in a disgust that masks the deep fear that an as-yet unknown, unthinkable intelligence might actively challenge us, preying on both the arrogant confidence in our own primacy and our fecklessness in the face of global disaster.

The fear and disgust aroused by teuthic invasions have been analysed with considerable wit by Vilém Flusser, who sums up an imagined squid–human encounter when he says, 'we necessarily want to experience *it*, too, and it necessarily wants to swallow *us*.'[43] Flusser is writing about *Vampyroteuthis infernalis*, which the official taxonomy has recently excluded from squids, setting it in an order all its own, despite the fact that its name continues to mean 'vampire squid from hell'. The infernal monster banned even from other teuthic portents, reminds us that Milton's anti-heroic traitor to existence would rather reign alone than serve in someone else's heaven.

In what he calls a fable, Flusser sets up the engagement between our two species by emphasizing the utter impossibility

Vampyroteuthis infernalis is one of the true bathyc monsters.

of it. 'Cephalopods are our antipodes', he says, and 'we are both banished from much of life's domain: it into the abyss, we onto the surfaces of the continents'.[44] This homologous banishment serves as the abyssal basis for Flusser's comments on what a consideration of the Vampire squid's existence by a human might mean, as well as – and this is where the irony of Flusser's delightful fable takes over – what an investigation of human knowledge would mean when 'made from the perspective of the vampyroteuthis'. To each of us, the other appears impossible, or at most a traitorous horror, because the antipodal environment in which each flourishes would be death to the other. Flusser ingeniously twists this mutual impossibility into a dialectical reconciliation by asking:

> How does the world look in which our negative model exists? Is it the same world as ours, only seen from a different perspective? Or is it an environment, with the vampyroteuthis at its center, that somewhere or somehow overlaps with our environment?[45]

These are some of the thorniest questions of our time, as they are the only kind of queries that can push us to try imagining our antipode, the extreme denial of our possibility.

When Henry Lee argued that giant squids were merely large versions of everyday calamaries, he hoped to shatter superstitious fears of the monstrous Kraken by making the apocalyptic beast into an ordinary squid. And he was right, as Kubodera proved once and for all. The ominous Kraken of old, as Lowes observed of Coleridge's monsters, has merged into the ordinary and classified squids of our present. Very un-Kraken-like squids appear in fish markets around the world, and are easily visible from pleasure boats and ferries. But for people like Donald Crowhurst who venture out to the deeper waters, or who have the misfortune of a Coleridgean imagination capable of conjuring vague threats looming in the abyssal dark, the watery dynamism of the Kraken continues to suggest something horrible. The power of squids to arouse the terror of an apocalyptic treason of creation endures as Krakenish Humboldts and vampire squids from hell recall the old questions of what manner of being they are, and why they should be showing themselves, and at this particular time.

As rumour, tall tale or legend, the Kraken holds no specific form enabling it to fit into a taxonomy, but it always plunges back into the murky depths: it is formless oceanic motion that defies attempts at categorical knowledge. In our scientific era, a direct encounter with Kraken can only be terrifying as the realization that the apocalypse is not a myth. The rumours, poems and stories about the Kraken – even those coming after Steenstrup – remind us of the anxiety we carry with us, though it sleeps silently beneath our cheery faith in science. One aspect of our anxiety over what the Kraken portends might be expressed as the possibility of our vast empire of knowledge coming to an end, or ceasing to matter as it is replaced by a hellishly antipodal science. Our view of ourselves as the primary beings with the greatest intelligence holds little merit to vampyric squids for whom communication takes the form of cannibalism. In his playful experiment of analysing the human condition from the 'negative utopia' of *V. infernalis*, Flusser imagines how easily human intelligence may be supplanted by another form so different from ours as to seem 'a repugnant horror'.[46] The awakening of the Kraken could cause us to learn that the terra firma we had mastered with our taxonomies was actually a beast that resembled a continent, and was now returning to the abyssal and lightless deeps. Could it be that we have been so enraptured by the beauty of Kubodera's shimmering *Architeuthis*, as to overlook what its wink might foreshadow?

In one of the best squid books of the past decade, Danna Staaf follows the aeons of cephalopod adaptation and evolution, compared to which 'the human timescale is evanescent', and observes that squids may be 'among the animals best able to adapt to a changing planet'.[47] Cephalopods have survived because of their kinetic adaptability, which suggests that they will endure beyond us. The playfulness with which Flusser contemplates the apocalypse continues in a quasi-documentary from the Discovery Channel, *The Future Is Wild* (2002), which plays on the squids' kinetic adaptability that Staaf finds compelling. In this series, biologists from different disciplines respond to the thought problem of imagining a world after the Anthropocene. Moving ahead incrementally, first 5 million years, then 100 million years, and finally 200 million years, the scientists describe the shifting global environment and the creatures that would be active. Stephen Palumbo from Harvard assures us in an early episode that 'all these organisms are plausible', given the projected climatological and geophysical conditions projected. And, he says,

From the mockumentary *The Future Is Wild* (2002), where marine biologist William Gilly imagines a future of squids colonizing the post-human land.

After humans disappear, squids could take over the forests as 'squibbons', and form bourgeois families – if only they can stop dying after sex.

'terrible environmental events', such as rising sea levels and continental drifts, would 'create mass extinctions'. But – and this is an important caveat – 'after a mass extinction, the sea would not remain empty long'. One of the organisms 'poised to take advantage of the demise of the fish' would be a 25 m (85 ft) 'rainbow squid' that changes colour to look like shoals of 'silver swimmers' – the descendants of plankton who would replace bony teleosts. The rainbow squid employs the strategy to lure flying predators into the water where it can eat them.

At this point, Monterey's noted marine biologist William Gilly explains the high level of intelligence that would enable squids to take over the world. The narrator asks, 'if squids are intelligent now, what would 200,000,000 years of evolution enable them to achieve?' One notable achievement, Gilly fantasizes, is that future squids 'will solve the problem of dying after spawning'. And with such extended lifespans, squids will begin to colonize the land, emptied by the mass extinctions. Boneless, squishy, intellectual squids – perhaps not unlike Cthulhu or Wyndham's Bathies – will develop compacted muscles sufficient to enable them to walk on their eight legs, using their tentacles to snatch food from the trees. This walking squid would, naturally enough, stand 5 m (16½ ft) tall and weigh 8 tons. Smaller squids would use their boneless flexibility to climb into the trees, where they could use their arms to swing from limb to limb like the gibbons of our era. In fact, these future tree-dwellers would be 'squibbons', who also develop their intelligence and longevity to begin forming social groups where the adults care for the young. Gilly concludes this thought experiment – and reveals a scientific wit as capable as that of any of his fable-spinning peers – by stating, 'if there is a community in the future, I believe it will be with cephalopods.' This speculative squid future charms us with the fantasy that squids will evolve into an image of what we imagine ourselves to be – a sustained community of caring families. In other words, squids will adapt into what is most alien to them – us.

The less charming, but more bracing thought experiment would be to imagine a future in which sea levels rise and the reigning intelligence belongs to a cannibalistic predator who thrives by adapting continually and even ruthlessly, and for whom the awakening of passion will only ever mean death. That is the future that belongs to squids.

Alice Shirley standing
in front of her work
*The Giant Squid (life
size)*, drawn from the
Natural History
Museum, London,
specimen in 2010
– in fresh squid ink.
The incomplete
specimen is over
8.5 m (28 ft) long;
the drawing slightly
wider.

Timeline of the Squid

BEFORE TIME BEGAN	450,000,000 YEARS AGO	280,000,000 YEARS AGO
The Kraken sleeps	Devonian period: First cephalopod species – *Plectroneceras cambria* – in the fossil record	*Phragmoteuthida*, the shelled common ancestor of *Decabrachia* and *Octobrachia*, appear

6,000,000 YEARS AGO	FOURTH CENTURY BCE	1554
Cambrian period: Cephalopoda appear in recognizable form	Aristotle sees squids in the Bay of Kalloni, and divides them into two types	Guillaume Rondelet depicts the *pisce monachi habitu* and *pisce episcopi habitu* in his *Libri piscibus marinis*

1845	1898	1986
Alcide d'Orbigny divides squids into the two sub-orders, Oegopsida and Myopsida	Carl Chun leads the Valdivia deep-sea expedition. Twelve years later he publishes his hand-coloured images of squids, including *Vampyroteuthis infernalis*, which he named	Malcolm Clarke publishes *Handbook for the Identification of Cephalopod Beaks*, making a meaningful distinction between squid species possible for the first time

180,000,000 YEARS AGO	150,000,000 YEARS AGO	65,000,000 TO 1,500,000 YEARS AGO

Phragmoteuthida go extinct, unshelled *Vampyromorphida* and the first creatures with a squid-like gladius instead of an external shell appear

Jurassic period: modern Coleoidea appear

Caenozoic period: Coleoidea develop behavioural repertoire still recognizable in squids today. Some squids begin to move back into shallow water

1639	1798	1831

Kraken beached on the coast of Iceland

Georges Cuvier invents the modern name of the class Cephalopoda

Alfred Tennyson publishes 'The Kraken'

1997	2012	200,000,000 YEARS FROM NOW	THE END OF TIME

Humboldt squids invade Monterey Bay

Tsunemi Kubodera films an *Architeuthis dux* in its natural habitat

Scientists conjecture that squids will move onto land, mate with gibbons and create sophisticated communities

The Kraken awakes

References

1 NATURAL HISTORIES FROM ARISTOTLE TO STEENSTRUP

1 Aristotle, *On Respiration*, in *Aristotle on Youth and Old Age, Life and Death, and Respiration*, trans. W. Ogle (London, 1897), 478a.

2 See Aristotle, *History of Animals*, trans. A. L. Peck (Cambridge, MA, 1965), note at 489a.

3 Aristotle, *Parts of Animals*, trans. A. L. Peck (Cambridge, MA, 1961), 654a.

4 Aristotle, *Generation of Animals*, trans. A. L. Peck (Cambridge, MA, 1963), 733a.

5 Aristotle, *History of Animals*, 549b.

6 Aristotle, *Parts of Animals*, 679.

7 Aristotle, *History of Animals*, 550b.

8 Henry George Liddell and Robert Scott, *A Greek–English Lexicon* (Oxford, 1968), p. 1783.

9 Aristotle, *Parts of Animals*, 654a.

10 Aristotle, *History of Animals*, 534a.

11 Clyde F. E. Roper and Elizabeth K. Shea note that 'only one published record of *Architeuthis* exists for the Mediterranean Sea', in 'Unanswered Questions about the Giant Squid *Architeuthis* (Architeuthidae) Illustrate our Incomplete Knowledge of Coleoid Cephalopods', *American Malacological Bulletin*, XXXI (2013), p. 111. Not to deny the possible identification altogether, however: conceivably Aristotle might have heard of a giant squid in conversations with sailors and merchants who had travelled to the western Mediterranean, and along the Atlantic coast of Spain, a known habitation of the modern *Architeuthis dux*.

12 D'Arcy Thompson, *A Glossary of Greek Fishes* (Oxford, 1947), p. 232.

13 Ibid., p. 260.

14 In Ambroise Paré, *On Monsters and Marvels*, trans. Janis L. Pallister (Chicago, IL, 1982), pp. 118–19.

15 Norman Douglas, *Birds and Beasts of the Greek Anthology* (New York, 1929), p. 149.

16 D'Arcy Wentworth Thompson, *On Aristotle as a Biologist, With a Prooemion on Herbert Spencer* (Oxford, 1913), p. 13. Thompson's deduction that Aristotle lived briefly on Lesbos has recently inspired two fanciful accounts of that time and of his procedure in writing the natural history. These are Armand Marie Leroi, *The Lagoon: How Aristotle Invented Science*, trans. Simon MacPherson (New York, 2014), and Susanna Gibson, *Animal, Vegetable, Mineral? How Eighteenth-century Science Disrupted the Natural Order* (Oxford, 2015).

17 Aristotle, *History of Animals*, 523b.

18 Nicholas Purcell, 'Eating Fish: The Paradoxes of Seafood', in *Food in Antiquity*, ed. John Wilkins, David Harvey and Mike Dobson (Exeter, 1995), p. 135.

19 Emily Vermeule, *Aspects of Death in Early Greek Art and Poetry* (Berkeley, CA, 1979), p. 185.

20 Ibid., p. 184.

21 Quoted by James Davidson, in 'Osophagia: Revolutionary Eating in Athens', in Wilkins et al., *Food in Antiquity*, p. 211.

22 Quoted in Jon Solomon, 'The Apician Sauce, *Ius Apicianum*', in Wilkins et al., *Food in Antiquity*, p. 124.

23 Maguelonne Toussaint-Samat, *The History of Food*, trans. Anthea Bell (Oxford, 1992), p. 306.

24 John K. Papadopoulos and Deborah Ruscillo, 'A *Ketos* in Early Athens: An Archaeology of Whales and Sea Monsters in the Greek World', *American Journal of Archaeology*, CVI (2002), p. 202.

25 Archistratus, *Fragments from The Life of Luxury*, trans. John Wilkins and Shaun Hill (London, 2011), p. 84, fragment 54.

26 Marcel Detienne and Jean-Pierre Vernant, *Cunning Intelligence in Greek Culture and Society*, trans. Janet Lloyd (Chicago, IL, 1991), p. 159.

27 Pliny, *Natural History*, trans. H. Rackham (Cambridge, MA, 1983), bk. 9, pp. 275, 199.

28 Ibid., p. 165.

29 Ibid., p. 219.

30 Thompson, *Greek Fishes*, p. 287.

31 Alan Davidson, *North Atlantic Seafood: A Comprehensive Guide with Recipes* (Berkeley, CA, 2003), p. 247.

32 Erich Pontoppidan, *The Natural History of Norway*, 2 vols (London, 1755), vol. II, pp. 177, 179, 183.

33 Pontoppidan, *History of Norway*, vol. II, pp. 183, 185.

34 Ibid., pp. iv, 186.

35 Bernard Heuvelmans, *The Kraken and the Colossal Octopus: In the Wake of Sea-monsters* (London, 2003), p. 153.

36 Quoted ibid., p. 156.

37 Lee, *Sea Monsters Unmasked*, p. 25.

38 The nineteenth-century naturalist Frank Buckland reports in a note that a correspondent wrote to say, 'I spent the best part of a day in St Malo, trying to find if the picture described by Montfort was in existence, and upon making enquiries of several persons who ought to know, besides going about myself, I found that such a picture was not known in St Malo. However, there appears to be such a story as Montfort speaks of known to the Malosien, and one person told me that he thought there was a picture at Marseilles painted to commemorate the fight said to have taken place between the octopus and a crew from St Malo. I might add that there is no such church or chapel called St. Thomas in St Malo' (*Log-book of a Fisherman and Zoologist* [London, 1876], p. 209n). According to Heuvelmans (*The Kraken*, p. 159n), 'Armand Landrin speaks of such an ex-voto which, according to him, was still in 1867 with the church of Notre Dame de la Garde, in Marseille: however, it commemorated an incident which had taken place off the coast of South Carolina.'

39 James Wilson (signed 'W'), 'Remarks on the Histories of the Kraken and Great Sea Serpent', *Blackwood's Edinburgh Magazine*, II/12

(March 1818), pp. 645–54, and III/13 (April 1818), pp. 33–43. Only a few years later, another Scotsman, Samuel Hibbert, recounted the everyday acceptance of the Kraken as fact among the people of the Shetland Isles, in *A Description of the Shetland Islands, Comprising an Account of their Geology, Scenery, Antiquities, and Superstitions* (Edinburgh, 1822; repr. Lerwick, 1891), pp. 260–61.

40 Wilson, 'Remarks on the Histories of the Kraken and Great Sea Serpent', *Blackwood's Edinburgh Magazine*, II/12 (March 1818), pp. 646, 646n, 651.

41 Japetus Steenstrup, 'On the Merman (Called the Sea Monk) Caught in the Øresund in the Time of King Christian III', trans. M Roeleveld, *Steenstrupia*, VI (1980), pp. 307, 309.

42 Roper and Shea, 'Unanswered Questions about the Giant Squid', p. 111. They rationalize the numerous appearances there by stating that they occurred in 'sites exposed to the North Atlantic'.

2 MODERN TEUTHOLOGY

1 Clyde F. E. Roper and Elizabeth K. Shea, 'Unanswered Questions about the Giant Squid *Archeteuthis* (Architeuthidae) Illustrate our Incomplete Knowledge of Coleoid Cephalopods', *American Malacological Bulletin*, XXXI (2013), p. 109.

2 Kir Nesis, *Cephalopods of the World: Squids, Cuttlefishes, Octopuses, and Allies*, trans. B. S. Levitov (Neptune City, NJ, 1982), p. 9.

3 Malcolm R. Clarke, 'Introduction', in *The Role of Cephalopods in the World's Oceans*, ed. Malcolm R. Clarke, *Philosophical Transactions of the Royal Society of London*, CCCLI (1996), p. 980.

4 Nesis, *Cephalopods of the World*, p. 9; Clarke, 'Introduction', p. 980.

5 Clarke, 'The Role of Cephalopods in the World's Oceans: General Conclusions and the Future', in *The Role of Cephalopods in the World's Oceans*, ed. Clarke, p. 1107.

6 Nesis, *Cephalopods of the World*, p. 9.

7   Danna Staaf, *Squid Empire: The Rise and Fall of Cephalopods* (Lebanon, NH, 2017), p. 104.

8   Gilbert L. Voss, 'Present Status and New Trends in Cephalopod Systematics', in *The Biology of Cephalopods*, ed. Marion Dixon and J. B. Messenger (*Symposia of the Zoological Society of London*, XXXVIII, 1977), p. 51.

9   Patrizia Jereb and Clyde F. E. Roper, eds, *Cephalopods of the World: An Annotated and Illustrated Catalogue of Cephalopod Species Known to Date* (Rome, 2010), p. 1.

10 Ibid., p. 10, p. 3.

11 Voss, 'Present Status and New Trends', pp. 55f, 52, 53, 54.

12 Malcolm R. Clarke, ed., *A Handbook for the Identification of Cephalopod Beaks* (Oxford, 1986), p. 2.

13 P. G. Rodhouse and Ch. M. Nigmatullin, 'The Role of Cephalopods as Consumers', in *The Role of Cephalopods in the World's Oceans*, ed. Clarke, p. 1004.

14 This appellation shows up in numerous cephalopod websites, such as *Tonmo: The Octopus Online Magazine* (www.tonmo.com) and Cephbase (http://cephbase.eol.org).

15 Roger T. Hanlon and John B. Messenger, *Cephalopod Behaviour* (Cambridge, 1996), p. 7.

16 Egbert Giles Leigh, *Adaptation and Diversity: Natural History and the Mathematics of Evolution* (San Francisco, CA, 1971), p. 76.

17 Mark Norman, *Cephalopods: A World Guide* (Hackenheim, 2000), p. 18.

18 Staaf, *Squid Empire*, p. 74.

19 Jereb and Roper, *Cephalopods of the World*, p. 37.

20 Ibid., p. 91.

21 Ibid., p. 97.

22 Ibid., p. 124; Norman, *Cephalopods: A World Guide*, p. 41.

23 Thomas Clements, Caitlin Colleary, Kenneth de Baets and Jakob Vinther, 'Buoyancy Mechanisms Limit Preservation of Coleoid Cephalopod Soft Tissues in Mesozoic Lagerstätten', *Paelontology*, LX/1 (2017), p. 9.

24 Jereb and Roper, *Cephalopods of the World*, p. 280.

25 Tia Ghose, 'Humboldt Squid Researchers Link Beachings, Mass "Suicides" To Poisonous Algae Blooms', www.huffingtonpost.com, 6 December 2017.

26 Anonymous, 'Large Aggregations of Purpleback Squid Attributed to Climate Change', www.thenews.com, 15 January 2017.

27 Daniel Jones, 'Huge Demand for Cuttlefish Is Making a Fortune for British Fishermen', www.thesun.co.uk, 25 September 2017.

28 Peter Boyle and Paul Rodhouse, *Cephalopods: Ecology and Fisheries* (Oxford, 2005).

29 Ibid., pp. 27–8.

30 Lydia M. Mäthger, Nadav Shashar and Roger T. Hanlon, 'Do Cephalopods Communicate Using Polarized Light Reflections from their Skin?', *Journal of Experimental Biology*, ccii (2009), pp. 2137–8; Boyle and Rodhouse, *Ecology and Fisheries*, p. 27; Christell Jozet-Alves, Anne-Sophie Darmaillacq and Jean G. Boal, 'Navigation in Cephalopods', in *Cephalopod Cognition*, ed. Anne-Sophie Darmaillacq, Ludovic Dickel and Jennifer Mather (Cambridge, 2014), p. 152.

31 Jozet-Alves et al., 'Navigation in Cephalopods', p. 153.

32 Boyle and Rodhouse, *Cephalopods*, p. 84.

33 Martin Moynihan, *Communication and Noncommunication by Cephalopods* (Bloomington, in, 1985), p. 22; Hanlon and Messenger, *Cephalopod Behaviour*, p. 14.

34 Jennifer A. Mather, Ulrike Griebel and Ruth A. Byrne, 'Squid Dances: An Ethogram of Postures and Actions of *Sepioteuthis Sepioidea* Squid with a Muscular Hydrostatic System', *Marine and Freshwater Behaviour and Physiology*, xliii (2010), pp. 45–61.

35 Ibid., p. 46.

36 Ibid., p. 49, fig. 2.

37 Hanlon and Messenger, *Cephalopod Behaviour*, p. 121.

38 Michael J. Kuba, Tamar Gutnick and Gordon M. Burghardt, 'Learning from Play in Octopus', in *Cephalopod Cognition*, ed. Darmaillacq et al., p. 61.

39 Jennifer A. Mather, 'Do Cephalopods have Pain and Suffering?', in *Animal Suffering: From Science to Law, An International Symposium*,

ed. Thierry Auffret van der Kemp and Martine Lachance (Toronto, 2013), p. 116.

40 Hanlon and Messenger, *Cephalopod Behaviour*, p. 10.

3 FOLK TALES AND LEGENDS

1 Richard Ellis, *Monsters of the Sea* (New York, 1994), p. 122.

2 Homer, *The Odyssey of Homer*, trans. Richmond Lattimore (Chicago, IL, 1967), bk. 12, pp. 81, 86–7, 257.

3 Aeschylus, *Agamemnon*, trans. Robert Fagles (New York, 1977), line 1244.

4 John Boardman, '"Very Like a Whale" – Classical Sea Monsters', in *Monsters and Demons in the Ancient and Medieval Worlds: Papers Presented in Honor of Edith Porada*, ed. Anne E. Farkas, Prudence O. Harper and Evelyn B. Harrison (Mainz, 1987), pp. 73, 78; Katharine Shepard, *The Fish-tailed Monster in Greek and Etruscan Art* (New York, 1940), pp. 43, 78; Emily Vermeule, *Aspects of Death in Early Greek Art and Poetry* (Berkeley, CA, 1979), pp. 186, 183.

5 Albert the Great, *Man and Beast* (*De animalibus, Books 22–26*), trans. James J. Scanlan (Binghamton, NY, 1987), pp. 334, 357, 373.

6 Quoted in Vicki Ellen Szabo, *Monstrous Fishes and the Mead-dark Sea: Whaling in the Medieval North Atlantic* (Boston, MA, 2008), p. 181.

7 Ibid., p. 181.

8 Quoted in Joseph Nigg, *Sea Monsters: A Voyage Around the World's Most Beguiling Map* (Chicago, IL, 2013), p. 145.

9 *Arrow Odd*, in *Seven Viking Romances*, trans. Hermann Pálsson and Paul Edwards (New York, 1985), p. 86.

10 Ibid., p. 86.

11 Fridtjof Nansen, *In Northern Mists: Arctic Exploration in Early Times*, trans. Arthur G. Chater, 2 vols (New York, 1911), vol. II, p. 234.

12 Nigg, *Sea Monsters*, p. 54. And see Henry George Liddell and Robert Scott, *A Greek–English Lexicon* (Oxford, 1968), art. κῆτος. Nigg's point is confirmed by John K. Papadopoulos and Deborah Ruscillo, 'A *Ketos* in Early Athens: An Archaeology of Whales and Sea

Monsters in the Greek World', *American Journal of Archaeology*, CVI (2002), p. 207.

13   Bernard Heuvelmans, *The Kraken and the Colossal Octopus: In the Wake of Sea-monsters* (London, 2003), p. 107.

14   Quoted in Nigg, *Sea Monsters*, p. 140.

15   Henry W. Lee, *Sea Monsters Unmasked and Sea Fables Explained* (London, 1883), pp. vi, 49.

16   Anthony Cornelis Oudemans, *The Great Sea-serpent: An Historical and Critical Treatise* (London, 1892), p. 105.

17   J. A. Teit, 'Water-beings in Shetlandic Folklore, as Remembered by Shetlanders in British Columbia', *Journal of American Folklore*, XXXI (1918), p. 197.

18   Oudemans, *The Great Sea-serpent*, p. 106.

19   Ellis, *Monsters of the Sea*, p. 85.

20   Frank Bullen, *The Cruise of the 'Cachalot': Round the World in Search of Sperm Whales* (Champaign, IL, 1903), p. 90.

21   Ibid., p. 91.

22   June Gutmanis, *Na Pule Kahiko: Ancient Hawaiian Prayers* (Honolulu, HI, 1983), p. 6.

23   Ibid., p. 97.

24   Ibid., p. 14.

25   Daren Kamali, 'Manteress: Transformation', in *Squid Out of Water: The Evolution* (Honolulu, HI, 2014) p. 9.

26   Ibid., Frontispiece.

27   Kamali, 'Three in One', ibid., p. 39.

28   Kamali, 'Desert Sun', ibid., p. 60.

29   Kamali, 'Octopast', ibid., pp. 41–2.

30   Jan Knappert, *Pacific Mythology: An Encyclopaedia of Myth and Legend* (London, 1992), art. 'Burotu', p. 38, art. 'Paradise', pp. 219–20.

31   Kamali, 'Na dua', in *Squid Out of Water*, p. 49.

32   Kamali, 'Nesian Ancestors', ibid., p. 54.

33   Kamali, 'Ink Fish Writes Again', ibid., p. 63.

34   Kamali, 'Sleep Swimming', ibid., p. 32.

35   Kamali, 'Say-Say', ibid., p. 37.

36 Andy Campbell, 'Squid Injects Woman's Tongue with Sperm Bag as She Eats in Korea', www.huffingtonpost.com, 15 June 2012.

37 G. M. Park, J. Y. Kim, J. H. Kim and J. K. Huh, 'Penetration of the Oral Mucosa by Parasite-like Sperm Bags of Squid: A Case Report in a Korean Woman', *Journal of Parasitology*, xcviii (February 2012), pp. 222–3.

38 See Daniel Engber, 'Rump Faker: Is Imitation Calamari Made from Pig Rectum? A Charming Urban Legend Gets Its Start', *Slate*, 13 January 2013.

39 Wendy Williams, *Kraken: The Curious, Exciting, and Slightly Disturbing Science of Squid* (New York, 2011), pp. 50, 53.

40 Ibid., pp. 55, 76, 77.

41 Luke Inman, www.youtube.com, 17 February 2010.

42 China Miéville, *Kraken: An Anatomy* (Basingstoke, 2010), p. 42.

43 Jeff VanderMeer, *City of Saints and Madmen* (Holicong, PA, 2002), p. 51.

44 A chart of some of these patterns appears as fig. 1 on p. 17 of Madnock, *King Squid*; this monograph is paginated separately from the narrative preceding.

45 Madnock, *King Squid*, pp. 16, 18.

4 KINETIC SQUID

1 A. L. Peck, 'Preface', in Aristotle's *Generation of Animals*, trans. Peck (Cambridge, MA, 1963), p. xlvi.

2 Ibid., p. lix.

3 Marion Nixon and John Z. Young, *The Brains and Lives of Cephalopods* (Oxford, 2003), p. 27.

4 J. P. Gilpin-Brown, 'The Squid and Its Giant Nerve Fibre', in *The Biology of Cephalopods*, ed. Marion Dixon and J. B. Messenger (*Symposia of the Zoological Society of London*, xxxviii, 1977), p. 235; Nixon and Young, *Brains and Lives*, p. 96.

5 Nixon and Young, *Brains and Lives*, p. 99.

6 Peter Boyle and Paul Rodhouse, *Cephalopods: Ecology and Fisheries* (Oxford, 2005), p. 28; Bernd U. Budelmann and Horst Bleckmann,

'A Lateral Line Analogue in Cephalopods: Water Waves Generate Microphonic Potentials in the Epidermal Head Lines of *Sepia* and *Lolliguncula*', *Journal of Comparative Physiology A*, CLXIV (1988), p. 4.

7 Patrizia Jereb and Clyde F. E. Roper, *Cephalopods of the World*, vol. II (Rome, 2010), p. 282.

8 Nixon and Young, *Brains and Lives*, p. 101; Boyle and Rodhouse, *Cephalopods*, p. 101.

9 Boyle and Rodhouse, *Cephalopods*, p. 21.

10 Ibid., p. 101.

11 Nixon and Young, *Brains and Lives*, p. 99.

12 Roger T. Hanlon, Michael R. Maxwell, Nadav Shashar, Ellis R. Loew and Kim-Laura Boyle, 'An Ethogram of Body Patterning Behavior in the Biomedically and Commercially Valuable Squid *Loligo pealei* off Cape Cod, Massachusetts', *Biological Bulletin*, CXCVII (1999), p. 51; Roger T. Hanlon and John B. Messenger, *Cephalopod Behaviour* (Cambridge, 1996), pp. 22, 24, 38, 39, 40.

13 Hanlon et al., 'An Ethogram of Body Patterning Behaviour', p. 51.

14 Jereb and Roper, *Cephalopods of the World*, pp. 64–5.

15 Hanlon et al., 'An Ethogram of Body Patterning Behaviour', p. 60.

16 Ibid., p. 51.

17 Ibid., p. 56.

18 Ibid., p. 58.

19 Ibid., p. 59.

20 Lydia M. Mäthger, Nadav Shashar and Roger T. Hanlon, 'Do Cephalopods Communicate Using Polarized Light Reflections from their Skin?', *Journal of Experimental Biology*, CCXII (2009), pp. 2133–4.

21 Ibid., p. 2134.

22 Hanlon and Messenger, *Cephalopod Behaviour*, pp. 164, 21.

23 John Field, 'Jumbo Squid (*Dosidicus gigas*) Invasions in the Eastern Pacific Ocean', *Symposium of the Californian Ocean Fisheries Investigations Conference*, XLIX (2008), p. 78; Louis D. Zeidberg and Bruce H. Robison, 'Invasive Range Expansion by the Humboldt Squid, *Dosidicus gigas*, in the Eastern North Pacific', *Proceedings of the National Academy of Science*, CIV (2007), p. 12949.

24 J. S. Stewart, E. L. Hazen, D. G. Foley, S. J. Bograd and W. F. Gilly, 'Marine Predator Migration during Range Expansion: Humboldt Squid *Dosidicus gigas*, in the Northern California Current System', *Marine Ecology Progress Series*, CDLXXI (2012), pp. 136, 146, 142.

25 Ibid., p. 147; Mark Denny, *How The Ocean Works: An Introduction to Oceanography* (Princeton, NJ, 2008), p. 166; Field, 'Jumbo Squid Invasions', p. 80; Brad A. Seibel, N. Sören Häfker, Katja Trübenbach, Jing Zhang, Shannon N. Tessier, Hans-Otto Pörtner, Rui Rosa and Kenneth B. Storey, 'Metabolic Suppression during Protracted Exposure to Hypoxia in the Jumbo Squid, *Dosidicus gigas*, Living in an Oxygen Minimum Zone', *Journal of Experimental Biology*, CCXVII (2014), p. 2556.

26 Seibel et al., 'Metabolic Suppression', p. 2555.

27 Paul G. Rodhouse, 'Large-scale Range Expansion and Variability in Ommastrephid Squid Populations: A Review of Environmental Links', *Symposium of the Californian Ocean Fisheries Investigation Conference*, XLIX (2008), p. 82.

28 Friedman Keyl, Juan Argüellas, Luís Maríategui, Ricardo Tafur, Matthias Wolff and Carmen Yamshiro, 'A Hypothesis on Range Expansion and Spatio-temporal Shifts in Size-at-maturity of Jumbo Squid (*Dosidicus gigas*) in the Eastern Pacific Ocean', *Symposium of the California Ocean Fisheries Investigation Conference*, XLIX (2008), p. 119.

29 Ibid., p. 121.

30 Ibid., pp. 119, 124.

31 Hanlon and Messenger, *Cephalopod Behaviour*, p. 159.

32 Ibid., p. 150.

33 The pattern of sex leading quickly to death is termed 'semelparous' breeding, from the mythic figure of Semele who died upon her one sexual encounter with Zeus.

34 R. A. Byrne, U. Griebel, J. B. Wood and J. A. Mather, 'Squid Say It with Skin: A Graphic Model for Skin Displays in Caribbean Reef Squid (*Sepioteuthis Sepioidea*)', in Proceedings of the International Symposium 'Coleoid Cephalopods Through Time', 17–19 September 2002, ed. K Warnke, H. Keupp and S. Boletzky, *Berliner Paläobiologische Abhandlungen*, III (2003), p. 29.

35 Denny similarly describes the interactions of decaying and sinking animals with ocean currents as a 'biological pump' and as 'an ineffective life-support system' (*How the Ocean Works*, pp. 166–7).

36 Nixon and Young, *Brains and Lives*, p. 97.

5 ANXIETY

1 Nicholas Tomalin and Ron Hall, *The Strange Last Voyage of Donald Crowhurst* (New York, 1970), p. 203 (ellipsis and emendations in the text).

2 Peter Godfrey-Smith, *Other Minds: The Octopus, the Sea, and the Deep Origins of Consciousness* (New York, 2016), p. 10.

3 John Keats, *Letters of John Keats: A Selection*, ed. Robert Gittings (Oxford, 1970), p. 237.

4 Southey did indeed believe that mermaids existed. In an 1814 review of James Forbes's *Oriental Memoirs*, Southey commented on the author's disappointment in not seeing one on his return journey to England. Forbes stated that their existence had been attested by Mr Matcham, superintendent of the East India Company's marine at Bombay, who 'frequently saw these animals, from six to twelve feet long; the head and face resembling the human, except that the nose and mouth rather more resembled the hog; the skin fine, and smooth; the neck, breast, and body of the female, as low as the hips, appeared, he said, like a well-formed woman; from thence to the extremity of the tail they were perfect fish. . . . These creatures, Mr. Matcham added, were regularly cut up and sold by weight in the fish markets at Mombaza' (*Quarterly Review*, XII, October 1814, p. 208).

5 Sir Walter Scott, *The Pirate*, ed. Mark A. Weinstein (Edinburgh, 2000), p. 18.

6 Sir Walter Scott, *Minstrelsy of the Scottish Border* (Edinburgh, 1873), vol. IV, p. 296n.

7 Samuel Taylor Coleridge, *Poetical Works*, ed. Ernest Hartley Coleridge (Oxford, 1912), p. 535.

8 John Livingston Lowes, *The Road to Xanadu: A Study in the Ways of the Imagination* (London, 1978), p. 111.

9 Alfred, Lord Tennyson, *The Poems of Tennyson*, ed. Christopher Ricks (London, 1969), pp. 246–7.

10 Christopher Ricks, *Tennyson* (New York, 1972), p. 44.

11 Edgar Allan Poe, 'MS Found in a Bottle', in *The Complete Tales and Poems* (New York, 1975), p. 118.

12 Ibid., pp. 121–2.

13 Ibid., p. 124.

14 Erich Pontoppidan, *Natural History of Norway*, 2 vols (London, 1755), vol. I, p. 85.

15 Poe, 'MS Found in a Bottle', p. 126.

16 Anthony Cornelius Oudemans, *The Great Sea-serpent: An Historical and Critical Treatise* (London, 1892), p. 105, quoting Olaus Magnus.

17 Herman Melville, *Moby Dick; or, The Whale* (Evanston, IL, 1988), p. 276.

18 Ibid., pp. 571, 572.

19 Victor Hugo, *Toilers of the Sea* (*Les Travailleurs de la mer*), trans. Sir Gilbert Campbell (Glasgow, n.d.), p. 271. The standard modern translation is that by James Hogarth (New York, 2002). I prefer Campbell's, for reasons stated in the text.

20 Ibid., p. 272.

21 Ibid., pp. 274–5.

22 Jules Verne, *Twenty Thousand Leagues under the Sea*, trans. Walter James Miller and Frederick Paul Walter (Annapolis, MD, 1993), p. 3.

23 Ibid., p. 4.

24 Ibid., p. 267.

25 Ibid., p. 352n.

26 Ibid., p. 348.

27 Ibid., p. 351.

28 Ibid., p. 352.

29 William Hope Hodgson, *The Boats of the 'Glen Carrig'* (New York, 1971), p. 48.

30 Ibid., pp. 48–9.

31 Ibid., p. 108.

32 Ibid., p. 119.

33 H. P. Lovecraft, 'The Call of Cthulhu', in *The Call of Cthulhu and Other Weird Stories*, ed. S. T. Joshi (New York, 1999), pp. 141, 163, 166, 168.

34 Ibid., p. 150.

35 Ibid., p. 153.

36 Ibid., p. 154.

37 Ibid., p. 153.

38 John Wyndham, *The Kraken Wakes* (New York, 1953), pp. 94–5.

39 Ibid., pp. 155, 46.

40 Ibid., p. 47.

41 Ibid., p. 51.

42 Ibid., pp. 229–30.

43 Vilém Flusser and Louis Bec, *Vampyroteuthis Infernalis: A Treatise, with a Report by the Institut Scientifique de Recherche Paranaturaliste*, trans. Valentine A. Pakis (Minneapolis, MN, 2012), p. 43.

44 Ibid., pp. 18, 25.

45 Ibid., pp. 70, 10, 30.

46 Ibid., p. 29.

47 Danna Staaf, *Squid Empire: The Rise and Fall of the Cephalopods* (Lebanon, NH, 2017), pp. 191, 190.

Select Bibliography

Albertus Magnus, *Man and the Beasts (De animalibus)*, trans. James J.
    Scanlan (Binghamton, NY, 1987)
Archestratus, *Fragments from The Life of Luxury*, trans. John Wilkins
    and Shaun Hill (Devon, 2011)
Aristotle, *Generation of Animals*, trans. A. L. Peck (Cambridge, MA, 1963)
—, *History of Animals*, trans. A. L. Peck (Cambridge, MA, 1965)
—, *On Respiration*, in *Aristotle on Youth and Old Age, Life and Death,
    and Respiration*, trans. W. Ogle (London, 1897)
—, *Parts of Animals*, trans. A. L. Peck (Cambridge, MA, 1961)
Arrow Odd, in *Seven Viking Romances*, trans. Hermann Pálsson and
    Paul Edwards (New York, 1985)
Boardman, John, '"Very Like a Whale" – Classical Sea Monsters', in
    *Monsters and Demons in the Ancient and Medieval Worlds: Papers
    Presented in Honor of Edith Porada*, ed. Anne E. Farkas, Prudence O.
    Harper and Evelyn B. Harrison (Mainz, 1987), pp. 73–84
Boyle, Peter, and Paul Rodhouse, *Cephalopods: Ecology and Fisheries*
    (Oxford, 2005)
Buckland, Frank, *Log-book of a Fisherman and Zoologist* (London, 1876)
Budelmann, Bernd U., and Horst Bleckmann, 'A Lateral Line Analogue
    in Cephalopods: Water Waves Generate Microphonic Potentials in
    the Epidermal Head Lines of *Sepia* and *Lolliguncula*', *Journal of
    Comparative Physiology A*, CLXIV (1988), pp. 1–5
Bullen, Frank, *The Cruise of the 'Cachalot': Round the World in Search of
    Sperm Whales* (Champaign, IL, 1903)
Byrne, R. A., U. Griebel, J. B. Wood and J. A. Mather, 'Squid Say It
    with Skin: A Graphic Model for Skin Displays in Caribbean Reef

Squid (*Sepioteuthis sepioidea*)', in Proceedings of the International Symposium 'Coleoid Cephalopods Through Time', 17–19 September 2002, ed. K Warnke, H. Keupp and S. Boletzky, *Berliner Paläobiologische Abhandlungen*, III (2003), pp. 29–35

Campbell, Andy, 'Squid Injects Woman's Tongue with Sperm Bag as She Eats in Korea', www.huffingtonpost.com, 15 June 2012

Canavan, Gerry, and Kim Stanley Robinson, *Green Planets: Ecology and Science Fiction* (Middletown, CT, 2014)

Clarke, Malcolm R., 'Introduction', in *The Role of Cephalopods in the World's Oceans*, ed. Malcolm R. Clarke, *Philosophical Transactions of the Royal Society of London*, CCCLI (1996), pp. 977–82

—, 'The Role of Cephalopods in the World's Oceans: General Conclusions and the Future', in *The Role of Cephalopods in the World's Oceans*, in *Philosophical Transactions of the Royal Society of London*, CCCLI (1996), pp. 1105–12

—, ed., *A Handbook for the Identification of Cephalopod Beaks* (Oxford, 1986)

—, *The Role of Cephalopods in the World's Oceans*, *Philosophical Transactions of the Royal Society of London*, CCCLI (1996), pp. 977–1112

Clements, Thomas, Caitlin Colleary, Kenneth de Baets and Jakob Vinther, 'Buoyancy Mechanisms Limit Preservation of Coleoid Cephalopod Soft Tissues in Mesozic Lagerstätten', *Paelontology*, LX/1 (2017), pp. 1–14

Coleridge, Samuel Taylor, *Poetical Works*, ed. Ernest Hartley Coleridge (Oxford, 1912)

Darmaillacq, Anne-Sophie, Ludovic Dickel and Jennifer Mather, eds, *Cephalopod Cognition* (Cambridge, 2014)

Davidson, Alan, *North Atlantic Seafood: A Comprehensive Guide with Recipes* (Berkeley, CA, 2003)

Davidson, James, 'Osophagia: Revolutionary Eating in Athens', in *Food in Antiquity*, ed. John Wilkins, David Harvey and Mike Dobson (Exeter, 1995), pp. 204–13

Denny, Mark, *How The Ocean Works: An Introduction to Oceanography* (Princeton, NJ, 2008)

Derrida, Jacques, *Points . . . Interviews, 1974–1994* (Stanford, CA, 1995)

Detienne, Marcel, and Jean-Pierre Vernant, *Cunning Intelligence in Greek Culture and Society*, trans. Janet Lloyd (Chicago, IL, 1991)

Dixon, Marion, and J. B. Messenger, eds, *The Biology of Cephalopods* (*Symposia of the Zoological Society of London*, XXXVIII, 1977)

Douglas, Norman, *Birds and Beasts of the Greek Anthology* (New York, 1929)

Ellis, Richard, *Monsters of the Sea* (New York, 1994)

—, *The Search for the Giant Squid: The Biology and Mythology of the World's Most Elusive Sea Creature* (New York, 1998)

Engber, Daniel, 'Rump Faker: Is Imitation Calamari Made from Pig Rectum? A Charming Urban Legend Gets Its Start', *Slate*, 13 January 2013

Field, John, 'Jumbo Squid (*Dosidicus gigas*) Invasions in the Eastern Pacific Ocean', *Symposium of the Californian Ocean Fisheries Investigations Conference*, XLIX (2008), pp. 79–81

Flusser, Vilém, and Louis Bec, *Vampyroteuthis Infernalis: A Treatise, with a Report by the Institut Scientifique de Recherche Paranaturaliste*, trans. Valentine A. Pakis (Minneapolis, MN, 2012)

Ghose, Tia, 'Humboldt Squid Researchers Link Beachings, Mass "Suicides" To Poisonous Algae Blooms', www.huffingtonpost.com, 6 December 2017

Gibson, Susannah, *Animal, Vegetable, Mineral? How Eighteenth-century Science Disrupted the Natural Order* (Oxford, 2015)

Gilpin-Brown, J. P., 'The Squid and Its Giant Nerve Fibre', in *The Biology of Cephalopods*, ed. Marion Dixon and J. B. Messenger (*Symposia of the Zoological Society of London*, XXXVIII, 1977), pp. 233–41

Godfrey-Smith, Peter, *Other Minds: The Octopus, the Sea, and the Deep Origins of Consciousness* (New York, 2016)

Gutmanis, June, *Na Pule Kahiko: Ancient Hawaiian Prayers* (Honolulu, HI, 1983)

Hanlon, Roger T., Michael R. Maxwell, Nadav Shashar, Ellis R. Loew and Kim-Laura Boyle, 'An Ethogram of Body Patterning Behavior in the Biomedically and Commercially Valuable Squid *Loligo Pealei* off Cape Cod, Massachusetts', *Biological Bulletin*, CXCVII (1999), pp. 49–62

Hanlon, Roger T., and John B. Messenger, *Cephalopod Behaviour* (Cambridge, 1996)

Heuvelmans, Bernard, *The Kraken and the Colossal Octopus: In the Wake of Sea-monsters* (London, 2003)

Hibbert, Samuel, *A Description of the Shetland Islands, Comprising an Account of their Geology, Scenery, Antiquities, and Superstitions* (Edinburgh, 1822; repr. Lewick, 1891)

Hodgson, William Hope, *The Boats of the 'Glen Carrig'* (New York, 1971)

Homer, *The Odyssey of Homer*, trans. Richmond Lattimore (New York, 1976)

Hugo, Victor, *Toilers of the Sea* (*Les Travailleurs de la mer*), trans. Sir Gilbert Campbell (Glasgow, n.d.)

Jereb, Patrizia, and Clyde F. E. Roper, eds, *Cephalopods of the World: An Annotated and Illustrated Catalogue of Cephalopod Species Known to Date* (Rome, 2010)

Jones, Daniel, 'Huge Demand for Cuttlefish is Making a Fortune for British Fishermen', www.thesun.co.uk, 25 September 2017

Jozet-Alves, Christell, Anne-Sophie Darmaillacq and Jean G. Boal, 'Navigation in Cephalopods', in *Cephalopod Cognition*, ed. Anne-Sophie Darmaillacq, Ludovic Dickel and Jennifer Mather (Cambridge, 2014), pp. 150–76

Kamali, Daren, *Squid Out of Water: The Evolution* (Honolulu, HI, 2014)

Keats, John, *Letters of John Keats: A Selection*, ed. Robert Gittings (Oxford, 1970)

Keyl, Friedman, Juan Argüellas, Luís Maríategui, Ricardo Tafur, Matthias Wolff and Carmen Yamshiro, 'A Hypothesis on Range Expansion and Spatio-temporal Shifts in Size-at-maturity of Jumbo Squid (*Dosidicus gigas*) in the Eastern Pacific Ocean', *Symposium of the California Ocean Fisheries Investigation Conference*, XLIX (2008), pp. 119–28

Knappert, Jan, *Pacific Mythology: An Encyclopaedia of Myth and Legend* (London, 1992)

Kuba, Michael J., Tamar Gutnick and Gordon M. Burghardt, 'Learning from Play in Octopus', in *Cephalopod Cognition*, ed. Anne-Sophie

Darmaillacq, Ludovic Dickel and Jennifer Mather (Cambridge, 2014), pp. 57–71

Lee, Henry W., *Sea Monsters Unmasked and Sea Fables Explained* (London, 1883)

Leigh, Egbert Giles, *Adaptation and Diversity: Natural History and the Mathematics of Evolution* (San Francisco, CA, 1971)

Lovecraft, H. P., *The Call of Cthulhu and Other Weird Stories*, ed. S. T. Joshi (New York, 1999)

Lowes, John Livingston, *The Road to Xanadu: A Study in the Ways of the Imagination* (London, 1978)

Magnus, Olaus, *Historia de gentibus septentrionalibus (Description of the Northern People)*, trans. Peter Fisher (London, 1996)

Masters, Ryan, 'The Vicious Giant *Dosidicus gigas*, the Humboldt Squid, Seems to Be Finding a New Home Right Off Our Shores', *Monterey Weekly News*, 10 March 2005

Mather, Jennifer A., 'Do Cephalopods have Pain and Suffering?', in *Animal Suffering: From Science to Law, An International Symposium*, ed. Thierry Auffret van der Kemp and Martine Lachance (Toronto, 2013), pp. 113–24

Mather, Jennifer A., Ulrike Griebel and Ruth A. Byrne, 'Squid Dances: An Ethogram of Postures and Actions of *Sepioteuthis sepioidea* Squid with a Muscular Hydrostatic System', *Marine and Freshwater Behaviour and Physiology*, XLIII (2010), pp. 45–61

Mäthger, Lydia M., Nadav Shashar and Roger T. Hanlon, 'Do Cephalopods Communicate Using Polarized Light Reflections from their Skin?', *Journal of Experimental Biology*, CCII (2009), pp. 2133–40

Melville, Herman, *Moby Dick; or, The Whale* (Evanston, IL, 1988)

Miéville, China, *Kraken: An Anatomy* (New York, 2011)

Moynihan, Martin, *Communication and Noncommunication by Cephalopods* (Bloomington, IN, 1985)

Nansen, Fridtjof, *In Northern Mists: Arctic Exploration in Early Times*, trans. Arthur G. Chater (New York, 1911)

Nessis, Kir, *Cephalopods of the World: Squids, Cuttlefishes, Octopuses, and Allies*, trans. B. S. Levitov (Neptune City, NJ, 1982)

Nigg, Joseph, *Sea Monsters: A Voyage around the World's Most Beguiling Map* (Chicago, 2013)

Nixon, Marion, and John Z. Young, *The Brains and Lives of Cephalopods* (Oxford, 2003)

Norman, Mark, *Cephalopods: A World Guide* (Hackenheim, 2000)

Oppian, *Halieutica*, in *Oppian, Colluthus, Tryphiodorus*, trans. A. W. Mair (Cambridge, MA, 1928)

Oudemans, Anthony Cornelis, *The Great Sea-serpent: An Historical and Critical Treatise* (London, 1892)

Palmer, Christopher, 'Ordinary Catastrophes: Paradoxes and Problems in Some Recent Post-apocalyptic Fictions', in *Green Planets: Ecology and Science Fiction*, ed. Gerry Canavan and Kim Stanley Robinson (Middletown, CT, 2014), pp. 158–75

Papadopoulos, John K., and Deborah Ruscillo, 'A *Ketos* in Early Athens: An Archaeology of Whales and Sea Monsters in the Greek World', *American Journal of Archaeology*, CVI (2002), pp. 187–227

Paré, Ambroise, *On Monsters and Marvels*, trans. Janis L. Pallister (Chicago, IL, 1982)

Park, G. M., J. Y. Kim, J. H. Kim and J. K. Huh, 'Penetration of the Oral Mucosa by Parasite-like Sperm Bags of Squid: A Case Report in a Korean Woman', *Journal of Parasitology*, XCVIII (February 2012), pp. 222–3

Pliny, *Natural History*, trans. H. Rackham (Cambridge, MA, 1983)

Poe, Edgar Allan, *The Complete Tales and Poems* (New York, 1975)

Pontoppidan, Erich, *The Natural History of Norway*, 2 vols (London, 1755)

Purcell, Nicholas, 'Eating Fish: The Paradoxes of Seafood', in *Food in Antiquity*, ed. John Wilkins, David Harvey and Mike Dobson (Exeter, 1995), pp. 132–49

Ricks, Christopher, *Tennyson* (New York, 1972)

Rodhouse, Paul G., 'Large-scale Range Expansion and Variability in Ommastrephid Squid Populations: A Review of Environmental Links', *Symposium of the Californian Ocean Fisheries Investigation Conference*, XLIX (2008), pp. 82–9

Rodhouse, P. G., and Ch. M. Nigmatullin, 'The Role of Cephalopods as Consumers', in *The Role of Cephalopods in the World's Oceans*, ed. Malcolm R. Clarke, *Philosophical Transactions of the Royal Society of London*, CCCLI (1996), pp. 1003–22

Roper, Clyde F. E., and Elizabeth K. Shea, 'Unanswered Questions about the Giant Squid *Architeuthis* (Architeuthidae) Illustrate our Incomplete Knowledge of Coleoid Cephalopods', *American Malacological Bulletin*, XXXI (2013), pp. 109–22

Salvador, Rodrigo B., and Barbara M. Tomotani, 'The Kraken: When Myth Encounters Science', *História, Ciências, Saúde – Manguinhos*, XXI (2014), pp. 971–94

Scott, Sir Walter, *Minstrelsy of the Scottish Border*, 4 vols (Edinburgh, 1873)

—, *The Pirate*, ed. Mark A. Weinstein (Edinburgh, 2000)

Seibel, Brad, et al., 'Metabolic Suppression during Protracted Exposure to Hypoxia in the Jumbo Squid, *Dosidicus Gigas*, Living in an Oxygen Minimum Zone', *Journal of Experimental Biology*, CCXVII (2014), pp. 2555–68

Shepard, Katharine, *The Fish-tailed Monster in Greek and Etruscan Art* (New York, 1940)

Solomon, Jon, 'The Apician Sauce, *Ius Apicianum*', in *Food in Antiquity*, ed. John Wilkins, David Harvey and Mike Dobson (Exeter, 1995), pp. 115–31

Southey, Robert, 'Review of James Forbes, *Oriental Memoirs*', *Quarterly Review*, XII (October 1814), pp. 180–227

Staaf, Danna, *Squid Empire: The Rise and Fall of the Cephalopods* (Lebanon, NH, 2017)

Steenstrup, Japetus, 'On the Merman (Called the Sea Monk) Caught in the Øresund in the Time of King Christian III', trans. M. Roeleveld, *Steenstrupia*, VI (1980), pp. 292–332

Stewart, J. S., E. L. Hazen, D. G. Foley, S. J. Bograd and W. F. Gilly, 'Marine Predator Migration during Range Expansion: Humboldt Squid *Dosidicus Gigas*, in the Northern California Current System', *Marine Ecology Progress Series*, CDLXXI (2012), pp. 135–50

Szabo, Vicki Ellen, *Monstrous Fishes and the Mead-dark Sea: Whaling in the Medieval North Atlantic* (Boston, MA, 2008)

Taylor, Jesse Oak, 'Tennyson's Elegy for the Anthropocene: Genre, Form, and Species Being', *Victorian Studies*, LVIII (2016), pp. 224–33

Teit, J. A., 'Water-beings in Shetlandic Folklore, as Remembered by Shetlanders in British Columbia', *Journal of American Folklore*, XXXI (1918), pp. 180–201

Tennyson, Alfred, Lord, *The Poems of Tennyson*, ed. Christopher Ricks (London, 1969)

Theophrastus, *On Winds and On Weather Signs*, trans. Jason G. Wood (London, 1894)

Thompson, D'Arcy Wentworth, *A Glossary of Greek Fishes* (Oxford, 1947)

Tomalin, Nicholas, and Ron Hall, *The Strange Last Voyage of Donald Crowhurst* (New York, 1970)

Toussaint-Samat, Maguelonne, *The History of Food*, trans. Anthea Bell (Oxford, 1992)

VanderMeer, Jeff, *City of Saints and Madmen* (Holicong, PA, 2002)

Vermeule, Emily, *Aspects of Death in Early Greek Art and Poetry* (Berkeley, CA, 1979)

Verne, Jules, *Twenty Thousand Leagues under the Sea*, trans. Walter James Miller and Frederick Paul Walter (Annapolis, MD, 1993)

Voss, Gilbert L., 'Present Status and New Trends in Cephalopod Systematics', in *The Biology of Cephalopods*, ed. Marion Dixon and J. B. Messenger (*Symposia of the Zoological Society of London*, XXXVIII, 1977), pp. 49–60

Williams, Wendy, *Kraken: The Curious, Exciting, and Slightly Disturbing Science of Squid* (New York, 2011)

Wilson, James, 'Remarks on the Histories of the Kraken and Great Sea Serpent', *Blackwood's Edinburgh Magazine*, II/12 (March 1818), pp. 645–54, and III/13 (April 1818), pp. 33–43

Wyndham, John, *The Kraken Wakes* (New York, 1953)

Zeidberg, Louis D., and Bruce H. Robison, 'Invasive Range Expansion by the Humboldt Squid, *Dosidicus gigas*, in the Eastern North Pacific', *Proceedings of the National Academy of Science*, 104 (2007), pp. 12948–50

Websites and Organizations

AUT LAB FOR CEPHALOPOD ECOLOGY AND SYSTEMATICS (ALCES)
www.aut.ac.nz/study/study-options/science/facilities/lab-for-cephalopod-ecology-and-systematics
Centred at the Auckland University for Technology (AUT), this laboratory investigates all cephalopods, but focuses on squids, in the New Zealand waters. Squid science at AUT began with Dr Steve O'Shea, who first captured paralarval (baby) giant squid and participated in the expedition that obtained the first *in situ* footage of live adult giant squid off Japan in 2012. The lab is currently run by Dr Kat Bolstad, a deep-sea squid biologist originally from the USA. In addition to her work at AUT, Dr Bolstad has participated in documentaries and dived in submersibles to depths of 1,000 m (3,300 ft) in the Antarctic to observe deep-sea cephalopods.

CEPHALOPOD INTERNATIONAL ADVISORY COUNCIL (CIAC)
https://cephalopoda.org
The foremost organization for the study of cephalopods and the dissemination of information about cephalopods. Founded in 1983, the aims of CIAC (according to their mission statement) are to stimulate, accelerate and influence the direction of cephalopod research, to provide help and advice on aspects of cephalopod biology, including those relevant to the management of the increasingly important cephalopod fisheries, and to spread information on past and current research. This site provides information about the organization, including upcoming CIAC conferences, and recent publications.

GEOGRAPHICAL INFORMATION SYSTEM FOR SQUID DISTRIBUTION IN
THE SOUTHERN OCEAN
www.nerc-bas.ac.uk/public/mlsd/squid-atlas
This site provides a list of the numerous species known to exist in the
southern hemisphere. The list includes references to research articles
establishing a species presence in an area.

THE GIANT SQUID: DRAGON OF THE DEEP
www.smithsonianmag.com/science-nature/the-giant-squid-dragon-
of-the-deep-18784038
An article by Riley Black recounting the development of knowledge of
Architeuthis dux since the nineteenth century. Useful for its illustrations
and general account.

HIGH SEAS ALLIANCE
http://highseasalliance.org
The High Seas Alliance is a partnership of organizations and groups
aimed at building a strong common voice and constituency for the conservation of the high seas. The objective of the Alliance is to facilitate
international cooperation to establish high seas protected areas and to
strengthen high seas governance.

IN SEARCH OF GIANT SQUID
https://seawifs.gsfc.nasa.gov/squid.html
Digital version of the Open Planet travelling exhibition, sponsored by
the Smithsonian. The various pages work as interactive exhibits to allow
visitors to explore the various aspects of squid biology and the cultural
responses to squids.

THE MARINE BIOLOGICAL LIBRARY AT WOODS HOLE
www.mbl.edu
Founded in 1888, Woods Hole fosters some of the most important
research in marine biology.

THE MEDITERRANEAN SCIENCE COMMISSION (CIESM)
http://ciesm.org/index.htm
Headquartered in Monaco, CIESM consists of 23 member states supporting a network of several thousand marine researchers. The aim is to understand, monitor and protect a fast-changing, highly impacted Mediterranean Sea. CIESM runs expert workshops, collaborative programmes and regular congresses, delivering authoritative, independent advice to national and international agencies.

MONTEREY BAY AQUARIUM AND MONTEREY BAY AQUARIUM RESEARCH INSTITUTE (MBARI)
www.montereybayaquarium.org
www.mbari.org
The aquarium and the research institute work together, the former providing public access to many of the species in Monterey Bay (including the infamous Humboldt squid) and the latter housing some of the top squid researchers in the world, including William Gilley and Julie Stewart. MBARI publishes the *Monterey Bay Aquarium Seafood Watch*, which provides detailed information about the various squid species that commonly show up in fish markets and restaurants. This invaluable document includes details about sustainability, and threats to the different species commonly fished.

NATIONAL RESOURCE CENTER FOR CEPHALOPODS (NRCC)
www.gulfbase.org/organization/national-resource-center-cephalopods
The NRCC is a scientific programme supporting the mission of the Marine Resources and Aquatic Technology Program (MRATP) of the Marine Biomedical Institute (MBI), University of Texas Medical Branch. Located in Galveston, Texas, NRCC offers production systems for the culture of cephalopods, dormitory-style housing for visiting scientists, office space, computing resources, dry and wet laboratories, aquaculture facilities and docking space for research vessels. Its new Harborside Campus resources include a state-of-the-art aquaculture facility in Ewing Hall. Established in 1975, the NRCC provides the biomedical research community with squid, cuttlefish and other cephalopods for research. It is the only facility

to date that has been able to culture squids from egg to adulthood for multiple generations.

OCEAN RECOVERY ALLIANCE (ORA)
www.oceanrecov.org
Founded by Douglas Woodring, the Ocean Recovery Alliance works in various ways to improve the oceanic environment. In particular the ORA has played an important role in the collection of plastic garbage from the ocean. It was recently awarded the Prince's Prize for Innovative Philanthropy by HSH Prince Albert II of Monaco.

POLICY-ORIENTED MARINE ENVIRONMENTAL RESEARCH FOR THE SOUTHERN EUROPEAN SEAS (PERSEUS)
www.perseus-net.eu
This is an ongoing research project that assesses the dual impact of human activity and natural pressures on the Mediterranean and Black Seas. PERSEUS merges natural and socio-economic sciences to predict the long-term effects of these pressures on marine ecosystems.

SAFINA CENTER
http://safinacenter.org
Founded with the help of a MacArthur Grant by Carl Safina, this centre, based at Stony Brook University in Long Island, New York, foregrounds its Sustainable Seafood Program to help consumers, chefs, retailers and the medical community discover the connection between human health, a healthy ocean, fishing and seafood.

SMITHSONIAN GIANT SQUID PAGE
https://ocean.si.edu/ocean-life/invertebrates/giant-squid
Maintained by Clyde Roper for the National Museum of Natural History, this site provides plenty of information about *Architeuthis dux*, along with links to other sites and videos associated with the giant squid.

SQUIDFISH.NET

www.squidfish.net

For those interested in fishing for squid. Includes information about tackle, and places to fish for squid. Also includes recipes for squid meals and snacks.

THE WATERMEN PROJECT

http://watermenproject.org

Based in Geneva, Switzerland, the Watermen Project has collected photos and videos of underwater expeditions to serve as a library of stories about marine life. Their hope is that these stories of encounters with large marine animals will motivate concerned people to help conserve the ocean habitats.

Acknowledgements

In 2015 Daniel Heath Justice (author of *Badger*) and Rachel Poliquin (*Beaver*) hosted a colloquium of Animal series authors, entitled *Animal Fest 2015*, at the University of British Columbia in Vancouver, BC. It was a delightful weekend of engaging with people who shared an interest in and care for animals, and it was in talking with my fellow Reaktion authors that I began thinking about squids. My thanks to Daniel and Rachel, and to all the other authors who reminded me of the excitement that comes with thinking about the other creatures with whom we share the world. Jonathan Burt proved once again to be an encouraging – and gently correcting – series editor, and I can only think of him warmly. Michael Leaman offered me plenty of guidance in the drafting of this book. Harry Gilonis remains one my favourite correspondents, and has me in his debt for all the work he put into the illustrations: someone get that man a beer! Charissa Prchal happily helped me with countless technical details through the years, on this project and others. Tim Murphy surprised me over and over again with his tentacular grasp of squid lore, and his appreciation for arcana squishy and weird. He is a valued colleague, and a cherished friend. My dear wife, Linda, read through various versions of all the chapters, and with her usual incisive intelligence kept me on the right path. And finally, my oldest mate, Kevin Jackson (*Moose*), to whom I dedicate this book, deserves thanks for introducing me to Jonathan, for helping me find the bottoms of a few bottles of wine, and for too many other gestures of friendship for me to name.

Photo Acknowledgements

The author and publishers wish to express their thanks to the below sources of illustrative material and/or permission to reproduce it. Some locations of artworks are also given below, in the interest of brevity:

Photo Alaska Fisheries Science Center, NOAA Fisheries: p. 52; from *The American Museum Journal*, VII/7 (November 1907): p. 58 (left); from *The Australasian Sketcher* (24 November 1877)/photo courtesy State Library Victoria, Melbourne: p. 104; Bibliothèque municipale de la ville de Laon (MS 422): p. 125; Bibliothèque nationale de France, Paris: p. 172; The British Museum, London: p. 82; Brown University Library, Providence, RI: p. 179; from Frank T. Bullen, *The Cruise of the 'Cachalot' Round the World after Sperm Whales*, 2nd edn (London, 1899): p. 103; from Carl Chun, *Die Cephalopoden*, vol. II – 'Myopsida, Octopoda' ('Wissenschaftliche Ergebnisse der Deutschen Tiefsee-Expedition auf dem Dampfer "Valdivia" 1898–1899') (Jena, 1915), photos courtesy MBLWHOI Library, MA: pp. 44, 51, 58 (right); from Samuel Taylor Coleridge, *Der alte Matrose* (Leipzig, 1877): p. 158; from Pierre Denys de Montfort, *Histoire naturelle, générale et particulière des mollusques*, vol. II (Paris, 1802): p. 38; from André-Étienne-Just-Pascal-Joseph-François d'Audebard de Férussac and Alcide Dessalines d'Orbigny, *Histoire naturelle, générale et particulière des céphalopodes acétabulifères vivants et fossiles*, vol. II – Atlas (Paris, 1835–48): pp. 10, 25, 55; from Giacomo Franco, *Descrittione geografica delle Isole, Città, & Fortezze principali, che si trovano in mare nel viaggio da Venetia a Costantinopoli* (Venice, 1597): p. 17; from Amanda M. Franklin, Zoe E. Squires and Devi Stuart-Fox, 'Does Predation Risk Affect Mating Behavior? An Experimental Test in Dumpling Squid (*Euprymna tasmanica*)', in *PLoS ONE*, IX/12, 31 December 2014/photo Zoe Squires: p. 63; from Conrad Gessner, *Fischbüch: das ist ein kurtze, doch vollkommne Beschreybung aller Fischen so in dem Meer und süssen Wasseren, Seen, Flüssen oder anderen Bächen* . . . (Zürich, 1563), photos courtesy The Linda Hall Library of Science, Engineering and Technology, Kansas City, MO, and Zentralbibliothek, Zürich: pp. 29, 32, 94 (top); from Augustus A. Gould, *Report on the Invertebrata of Massachusetts*, 2nd edn (Boston, MA, 1870): p. 26; Michael Greenfelder/Alamy Stock Photo: p. 128; Henry-Dunant-Museum, Heiden (Switzerland): p. 169; from Isidore of Seville, *Isidori iunioris Hispalensis episcopi prologus in librum de responsione mundi & astrorum ordinatione* (Augsburg, 1472): p. 124; photo Andrea Izzotti/ Shutterstock.com: p. 6; J. Paul Getty Museum, Los Angeles: p. 91; James Ford Bell Library, University of Minnesota, Minneapolis: pp. 93, 94 (foot), 164; from Giuseppe Jatta, *I Cephalopodi viventi nel Golfo di Napoli* (Berlin, 1896): p. 14; photo courtesy Det Kongelige Bibliotek, Copenhagen: p. 40; from Henry Lee, *Sea Monsters Unmasked* (London, 1883): pp. 96, 97; reproduced by kind permission of Shannon MacGregor: p. 160; from Olaus Magnus, *Historiae de gentibus septentrionalibus* (Antwerp, 1557), photo courtesy Smithsonian Libraries, Washington, DC: p. 89; from Alfred Goldsborough Mayer, *Sea-shore Life: The Invertebrates of the New York Coast and the Adjacent Coast Region* (New York, 1905): p. 152; The Metropolitan Museum of Art, New York: pp. 22, 84, 85; Musée du quai Branly – Jacques Chirac, Paris/photo Francesco Bini (Sailko): p. 107; Museo Archeologico del Territorio di Populonia, Piombino/photo Wolfgang Sauber: p. 24 (foot); Museu Nacional Arqueològic de Tarragona (MNAT): p. 24 (top); Museu Nacional d'Art de Catalunya, Barcelona: p. 21; Museum of Natural History, University of Michigan, Ann Arbor/ photo Eric Christensen: p. 59; from Bent J. Muus, *Skallus, Søtænder, Blæksprutter* ('Danmarks Fauna', vol. LXV) (Copenhagen, 1959): p. 54; National Institute of Water and Atmospheric Research (NIWA), Wellington: p. 56; photo NOAA Office of Ocean Exploration and Research, Gulf of Mexico 2012 Expedition: p. 138; from Alcide Dessalines d'Orbigny, *Mollusques vivants et fossiles, ou, description de toutes les espèces de coquilles et de mollusques*, vol. I – Atlas (Paris, 1845), photo courtesy Ernst Mayr Library, Museum of Comparative Zoology, Harvard University, Cambridge, MA:

p. 53; and Zoe Squires, the copyright holder of the image on p. 63, have published them online under conditions imposed by a Creative Commons Attribution-ShareAlike 4.0 International License.

Readers are free to:

share – copy and redistribute the material in any medium or format.
adapt – remix, transform, and build upon the material for any purpose, even commercially.

Under the following terms:

attribution – You must give appropriate credit, provide a link to the license, and indicate if changes were made. You may do so in any reasonable manner, but not in any way that suggests the licensor endorses you or your use.
share alike – If you remix, transform, or build upon the material, you must distribute your contributions under the same license as the original.

Index

Page numbers in *italics* indicate illustrations